STAND

Mathematical Eye

Adam Hart-Davis

illustrated by Jane Cope

UNWIN HYMAN

Published in 1989 by
UNWIN HYMAN LIMITED
15/17 Broadwick Street
London W1V 1FP

British Cataloguing in Publication Data

Hart-Davis, Adam
Mathematical eye.
1. Mathematics
1. Title
510

ISBN 0 4448 0431

Designed by Geoffrey Wadsley
Typeset by Cambridge Photosetting Services
Origination by Chroma Graphics Singapore
Printed and bound in Great Britain

Preface

Mathematics is beautiful, elegant, and fun. But alas many pupils find it hard, boring, and irrelevant to their world. Why? Perhaps if it is to be enjoyed, maths has to be discovered, rather than rammed down your throat.

In a perfect world, maths should be inspirational. When you can find mathematical ideas for yourself you are delighted; you are excited; you will remember. So I have tried in this book to set up scenes and stories that will allow readers to find things out for themselves.

Mathematics is an abstract art – a constructional toy, glued together with logic. It needs no reality. But most ordinary people need their maths to cope with the real world of food and buses and money. So I have set most of the book in the real world, or in little scenes of fantasy that relate to the real world. Few readers will believe in Dracula's Cafe, or even in Hengist's Castle. But the ideas are real enough.

I started this book just when GCSE was arriving. The National Curriculum provided another source of true wisdom and ideal goals. The style of the book follows Cockcroft recommendations; the content is in line with the National Curriculum.

I started producing the television series roughly when the first draft of this book was complete. There is an approximate one-to-one correspondence between the two. So there is a programme on Circles and a chapter on Circles; a programme on Graphs and a chapter on Graphs. Sometimes I have used the same ideas on the screen as on the page, when they seemed to work well in both places. At other times the book and the series diverge, to provide a greater variation in ideas and practice. The book may be used in conjunction with the series or on its own.

The real reason I started on this book was that my younger son Jason complained that all his maths textbooks were so boring. I hope this one proves to be an exception: useful for teachers, and entertaining for pupils.

My grateful thanks for their help to Phil Cottier, Maureen Yeatman, John Walker, Simon Welfare, David Wells, and the Embassy of Finland.

Adam Hart-Davis

Contents

Hengist's castle

In about AD 500, when Alfred was ruling Mercia, a tough Saxon warrior called Hengist came to England to seek his fortune. He stopped a few riots, and did some other favours for a war-lord called Vortigern; so Vortigern owed him a favour.

Vortigern was delighted. He thought he could repay the debt by giving Hengist a tiny ***area*** *of land—hardly enough to stand on.*

But Hengist was cunning. First he found and killed an enormous ox. He skinned it and cured the hide. Then he worked out a way of cutting the hide into one very narrow strip. With great care and a sharp knife he worked away, never cutting through the strip, but making it longer and longer.

He had to use the strip as his ***perimeter****, and he knew that the longer he could make the perimeter, the larger the area of land he could claim.*

When he had finished, the strip was still a single piece of leather, but it was so amazingly thin that it made a huge perimeter—long enough to stretch all the way round a great hill.

Vortigern had to keep his word. Hengist got the hill, and on it he built a huge castle. The Saxons had arrived in England.

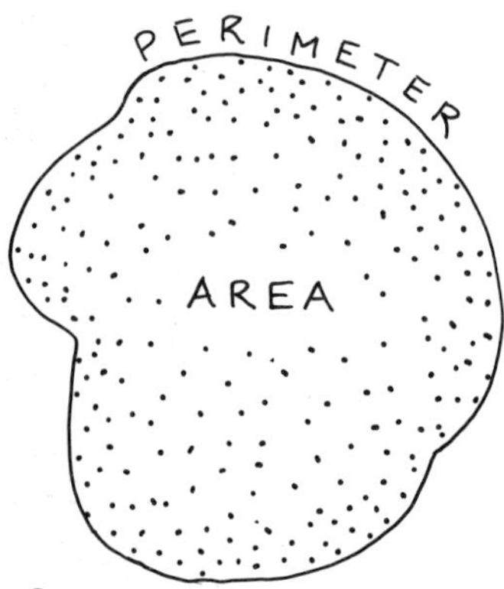

Remember

Perimeter *is the distance all the way round the edge or the sides of a shape. Perimeters can be measured in metres (m).*

Area *is the surface measurement of the shape inside the perimeter. Areas can be measured in square metres (sq m).*

Suppose Hengist's thong was 300 metres long. How big an area of land could he have enclosed? Do different shapes with the same perimeter have different areas? What shape would have been best for Hengist?

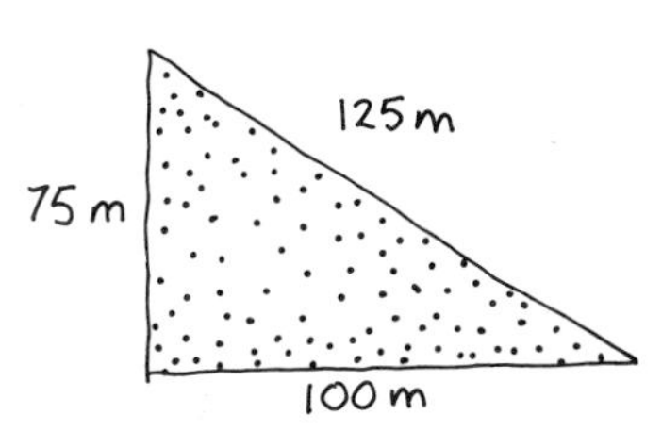

① He might have chosen a right-angled triangle, with sides of 75 m, 100 m, and 125 m. What would its area be? (Hint: half base × height.)

Hengist might have made a square.

② How long would the sides have been? (Perimeter = 300 m) Would the area of this square be bigger than that of the triangle?

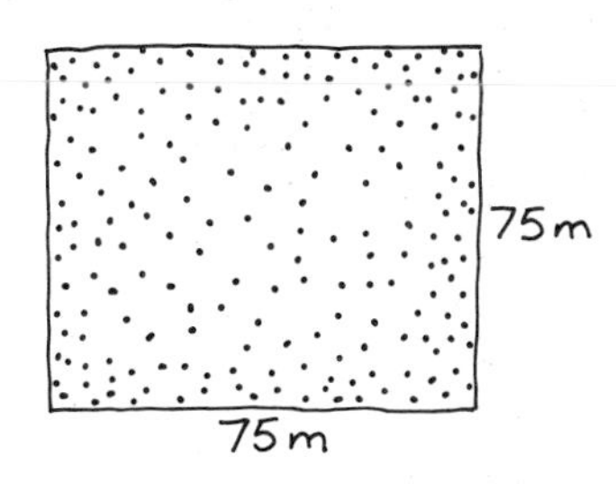

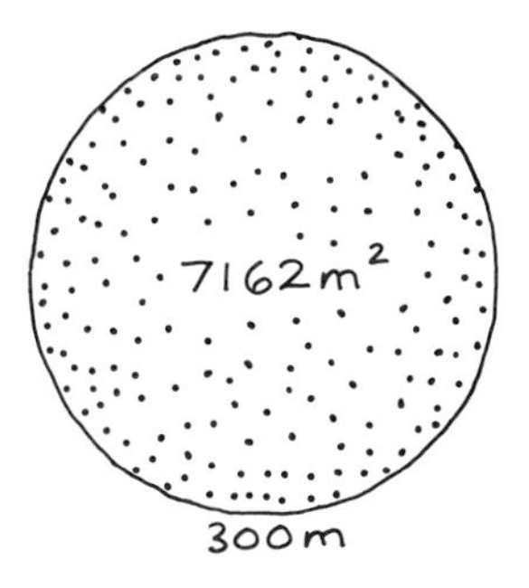

When your perimeter is fixed, you can enclose the maximum area by making a circle. Hengist was bright; so he probably worked this out. If he had made a circle with a circumference of 300 m, the area inside would have been about 7162 square metres.

③ How much bigger is this than the area of the square above?

④ Suppose you had just 1 metre of smart ribbon to make the border of a poster. You could make the poster a pentagon, a circle, a triangle, a hexagon, or a square. Which shape will give you the biggest poster? List the other shapes in order of decreasing size.

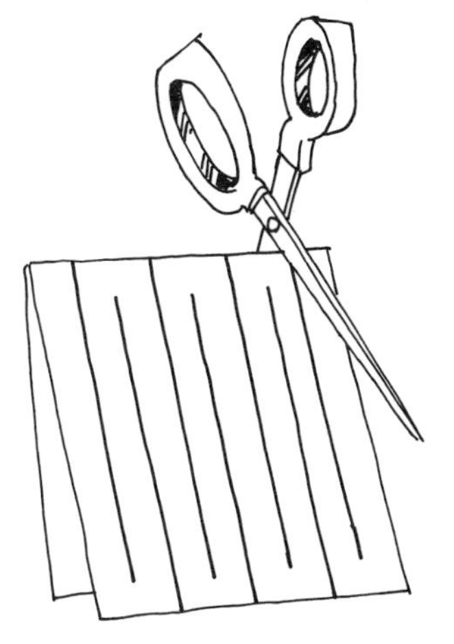

⑤ Hengist cut up an oxhide to give a long thong. Try the same thing with a piece of paper. Make a perimeter long enough to allow you to get through the hole. Can you get through a piece of paper the size of a postcard (15 cm × 10 cm) if you cut it up in the right way?

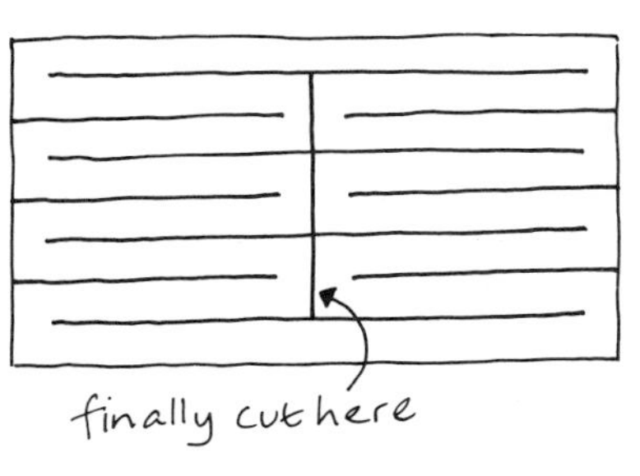

Here is one way you could cut the paper. Pull it out carefully to make one big loop. How big a loop can you make? Can you find a better pattern of cutting, to give a bigger loop? How small a piece of paper can you get through?

Belinda's brainwave

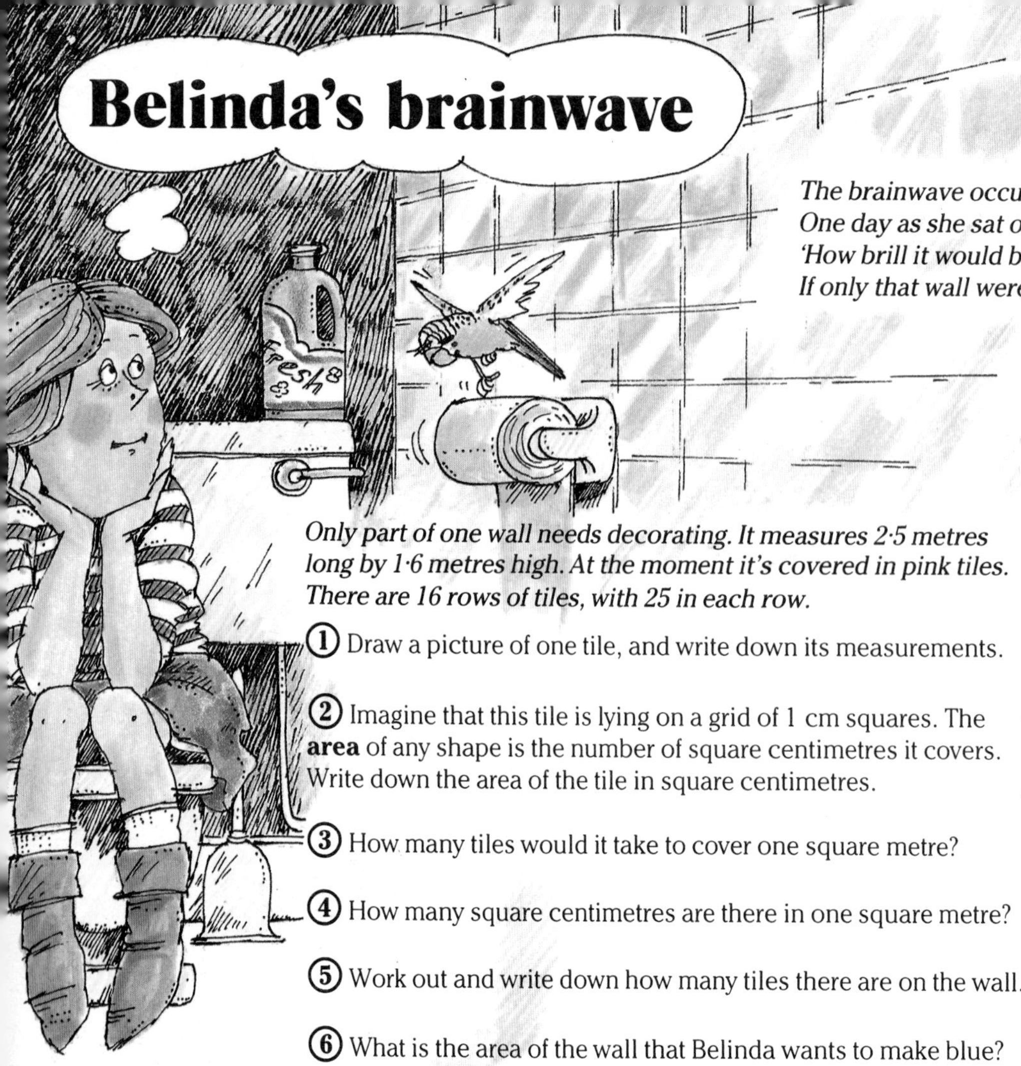

The brainwave occured to Belinda
One day as she sat on the loo.
'How brill it would be for my budgie and me
If only that wall were bright blue.'

Only part of one wall needs decorating. It measures 2·5 metres long by 1·6 metres high. At the moment it's covered in pink tiles. There are 16 rows of tiles, with 25 in each row.

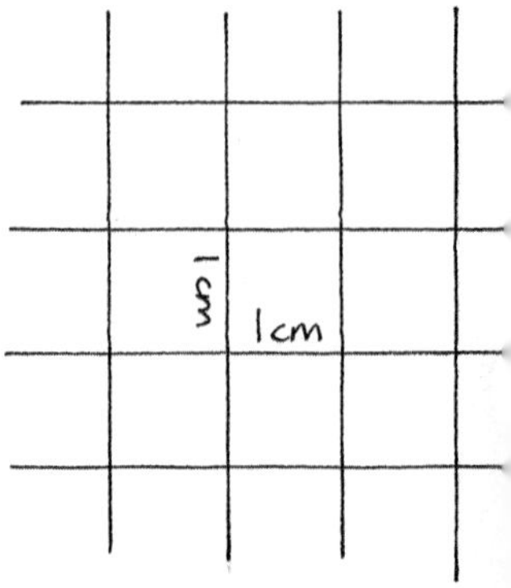

① Draw a picture of one tile, and write down its measurements.

② Imagine that this tile is lying on a grid of 1 cm squares. The **area** of any shape is the number of square centimetres it covers. Write down the area of the tile in square centimetres.

1 cm
1 cm

③ How many tiles would it take to cover one square metre?

④ How many square centimetres are there in one square metre?

⑤ Work out and write down how many tiles there are on the wall.

⑥ What is the area of the wall that Belinda wants to make blue?

Blue tiles of the same size as the pink ones cost £20 per 100.

The cement to stick them on comes in tubes. Each tube costs £1, and covers 1 square metre. Taking off the old tiles would be lots of hard work for Belinda, but wouldn't cost her anything.

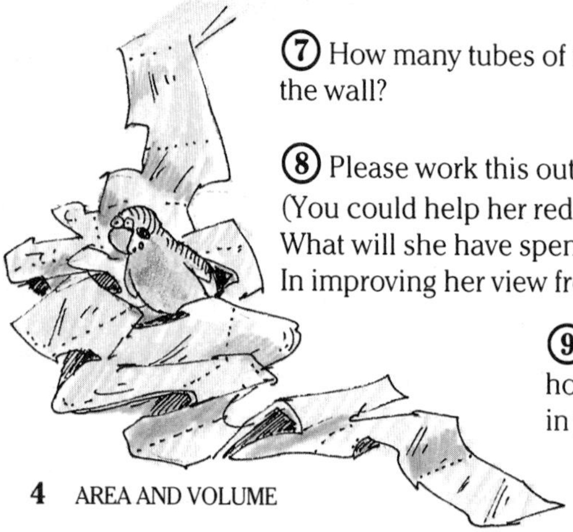

⑦ How many tubes of tile cement would she need for the wall?

⑧ Please work this out for Belinda.
(You could help her redecorate too!)—
What will she have spent, for the tiles and cement,
In improving her view from the loo?

Tile Cement

⑨ Find out about real bathroom tiles. How big are they? About how many—and how much cement—would you need for one wall in your bathroom? How much would it all cost?

Somebody's lunch

Something has been eating this leaf. Green salad for lunch.

① Write down the area of this page, in square centimetres. (Hint: Each square is 1 sq cm. How many squares are there altogether?)

② Estimate the area of the whole leaf, before anything chewed it. (Hint: How many squares would the whole leaf have covered?)

③ Estimate the area of leaf that has not been eaten.

④ The leaf was 0·2 mm thick. What volume of leaf has been eaten?

⑤ About what fraction of the leaf has been eaten?

⑥ About what percentage of the leaf has been eaten? (A calculator will be useful.)

There's nothing I like better than a crunchy leaf for lunch. Guzzle, guzzle, guzzle, munch, munch munch...

⑦ Whose lunch do you think the leaf has become?

A million cubes of sugar

There is no such event as the National Sugar Championship. But suppose there was, you might need to learn a lot about sugar lumps. Here is your chance.

A small sugar cube is 1 cm × 1 cm × 1 cm. It's a 1-centimetre cube, and its volume is 1 cubic centimetre.

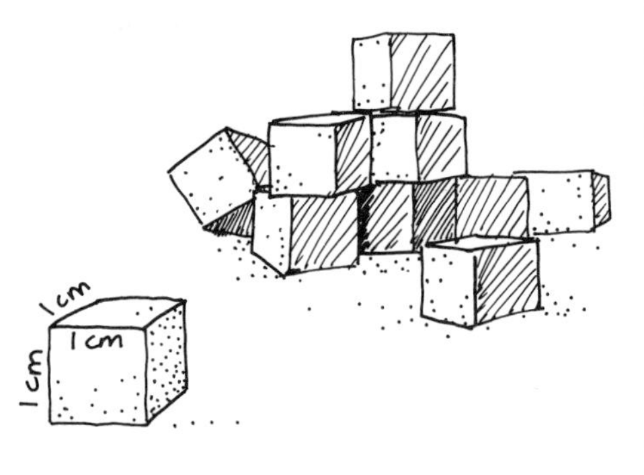

This box of sugar cubes is

10 cm long; so 10 cubes fit along the inside.
8 cm wide; so 8 cubes fit from front to back.
5 cm deep—enough to take 5 layers of cubes.

① How many cubes are there in the top layer?

② What is the area of the top layer? Remember, it is 10 cm long and 8 cm wide.

③ How many layers are there in the box?

④ How many cubes are there in the whole box?

⑤ What is the volume of the box in cubic centimetres?

If 400 cubes fill a little cardboard box, think about how much space you would need for a million. Suppose you won the National Sugar Championship, and your prize was a million cubes. Where could you put them?

⑥ Laid out in a line, would a million sugar cubes stretch through the channel tunnel (35 km long)?

⑦ Stacked one on top of another, would a million cubes of sugar make a pile as high as Mt Everest (8890 m)?

⑧ Spread out in a square, one layer thick, which of these would they cover? A classroom floor? A tennis court? A football field? London? Wales?

⑨ What is their volume, in cubic metres? How long is each edge of the cube-shaped crate that would just contain them all?

Mad Mex

The picture shows a mad Mexican riding a bicycle.

Sometimes he tries riding the burro and leading the bike. But that doesn't work well; the bike keeps falling over.

① What shape is the hat?

② The Mexican's bicycle is just over two metres long. About how wide is his hat?

③ Measure the distance from the exact centre of the hat to the brim in four different directions—up, down, left, and right. Is the distance the same in each direction?

The distance from the centre of a circle to the edge is the same in all directions. This distance is called the radius.

④ Copy and complete this sentence:

A circle is the path of a moving point which always stays the same ________ from a fixed point. This fixed point is called the ______ of the circle.

⑤ The spokes of Mex's bike are all the same length. Why do you think this is so? What would happen if they were all different?

⑥ When he stops for the night, the mad Mexican goes to sleep under his hat. He ties the end of the burro's string to a peg in the middle of a grass field. During the night, the burro eats all the grass it can reach.

What is the shape of the patch of short grass in the morning? Would it be a different shape if the string was longer, or shorter? Explain how you know the shape.

⑦ Which two words are 18 cm from the top left corner of this page and 15 cm from the bottom right corner? How did you find out?

⑧ Mad Mex has a rectangular vegetable patch, 4 m long and 2 m wide. He wants to plant as many beans in it as possible. The packet says he should plant them 20 cm apart, but what pattern should he use? Investigate various patterns of planting a fixed area.

Write a note to Mad Mex. Explain what he should do, with a diagram to show the best pattern for the maximum number of beans, and instructions on how to get all the beans the right distance apart.

Mad Mex 2

Mad Mex needs a new piece of ribbon to sew round the brim of his hat. How long a piece does he need?

A straight line joining two points on a circle and passing through the centre is called a **diameter**. *A straight line from the centre to any point on the circle is a* **radius**. *The distance all the way round a circle is the* **circumference**.

Mex's hat has a diameter of 2 metres. What he needs to know is its circumference.

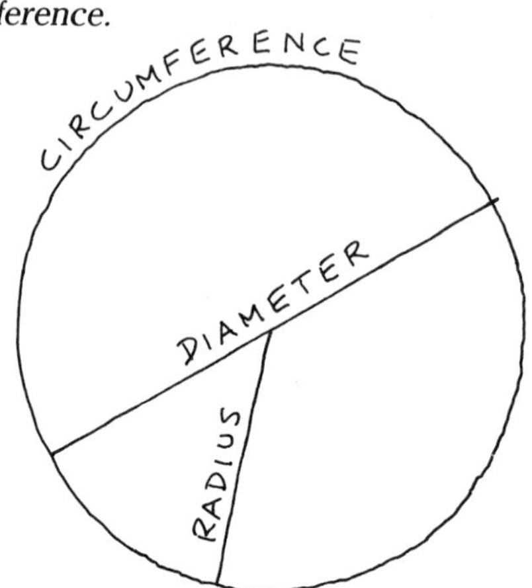

① Use a ruler to measure the diameter of each circle on this page and on pages 8 and 9. Measure round each circumference using a piece of paper or string.

Record your measurements in a table like this. Then, for each circle, divide the circumference by the diameter, and write the result in your table.

Circle	Diameter	Circumference	$\frac{\text{Circumference}}{\text{Diameter}}$
Hat	7cm	22cm	3.1

② Do the same for some real circles—plates, mugs, or coins. Or make some tubes by loosely rolling up pieces of paper. For each one, measure the circumference and the diameter, and record the results.

③ If you could drill through the centre of the Earth, the tunnel would be about 12 700 km long. The equator is about 40 000 km long. Write the diameter and circumference of the Earth in your table.

What is $\frac{\text{circumference}}{\text{diameter}}$ for the Earth?

④ What do you notice about the ratio $\frac{\text{circumference}}{\text{diameter}}$ for all the different circles?

The results you get for all the circles should be nearly the same—just over 3. The ancient Greeks found out that accurate circles, carefully measured, always gave the same value for $\frac{\text{circumference}}{\text{diameter}}$. They gave this value the nickname π (pi), one of the letters of their alphabet.

Measured carefully, π turns out to be about $3\frac{1}{7}$ —i.e. $\frac{22}{7}$ —or 3·14. These values are approximations, but near enough for everyday use. A more accurate value is 3·14159, and an even more accurate one is 3·141 592 653 589 793 238 462 643 383 279 502 . . .

Using computers, people have worked out π to 134 million decimal places, but still nobody knows its exact value. π is a number that cannot be written down completely, because it goes on for ever.

⑤ Mad Mex takes 15 minutes to cycle round the circular football stadium at a steady speed. How long do you think it would take him to bike straight across it at the same speed? (Use $\pi = 3$)

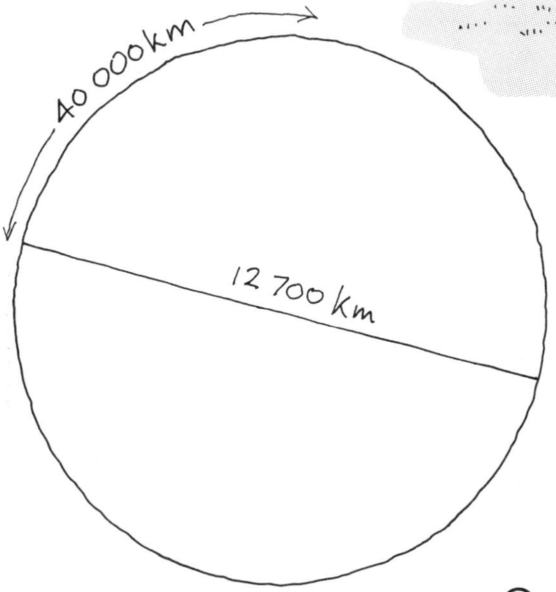

40 000 km

12 700 km

30cm

⑥ If the spokes of Mad Mex's bike are 30 cm long, and the tyres are 5 cm thick, how far does he go each time the wheels turn round once? (Use $\pi = 3\frac{1}{7}$)

⑦ How long is the brim of Mad Mex's hat?

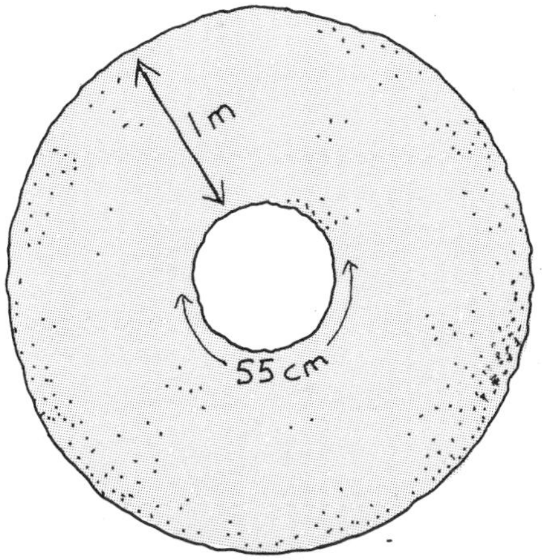

⑧ Mad Mex wants to make a new poncho from a circular piece of cloth. He wants it to be 1 metre long, and he wants a circular hole in the middle that he can just get his head through. He knows the hole must have a circumference of 55 cm, because that is his hat size. Above all, he doesn't want any seams; so it has to be made from one piece of material. But he doesn't know where to start.

Write him a set of instructions to explain (a) what width of material he needs to start with, (b) how to find the centre, (c) how to draw the circles, (d) how big a circle to draw for the whole poncho, and (e) how big a circle to draw for the hole.

1m

55 cm

Rope around the world

Here's a puzzle to baffle your friends.

Suppose
a rope was tied
all the way round the world,
from where you are now all the way back
to where you are now. 40 000 kilometres of rope!
You would be sure to find a piece of rope left over.
Suppose there turned out to be just twelve metres spare.

To use up the extra 12 metres, imagine the rope was lifted up, off the ground, to the same height all around the world.

How high could it be lifted? That is the question. If it were raised the same amount all around the world, how high would it be? Could you slide a piece of paper under it? Could you crawl under it, or walk under it? What do you think?

12m spare

Hint: imagine a rope wrapped round a ball with a diameter of 1 metre. That would take about 3 metres of rope. How much extra would be needed to lift the rope 1 metre off the ball, all the way round? How much extra would be needed to lift the rope 2 m? What about 3 m?

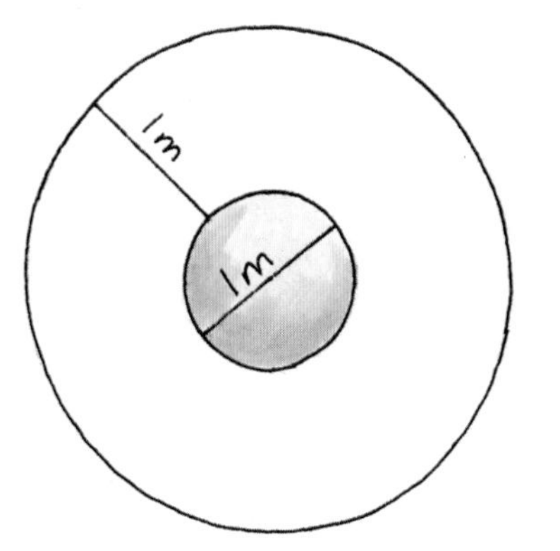

Then imagine the rope wrapped round a 2 metre ball. How much rope would you need, and how much extra to raise that by 1 m, and by 2 m? What about some larger balls?

The answer is that you could lift the rope nearly 2 metres to take up the 12 spare metres. The size of the ball makes no difference at all. An extra 12 m of circumference means an extra 4 m of diameter, or an extra 2 m of radius. So even if the rope went all the way round the world, when you took up the extra 12 m, you could walk underneath it.

Circles within circles

Investigate patterns you can make by drawing circles with a pair of compasses on large pieces of paper. Here are three ideas to start with.

① (a) Draw a circle.

(b) Do not alter your compasses. Pick a point on the circumference as the next centre, and draw another circle the same size.

(c) Still keep your compasses the same. Choose for your next centre one of the points where the two circles cross. Draw another circle.

(d) Keep doing this until the pattern is clear.

You could colour in a few of the segments to pick out some of the repeated shapes. Or look for regular hexagons or other polygons.

If you have time, try a different pattern, by making the first circle half as big, or twice as big, as the others.

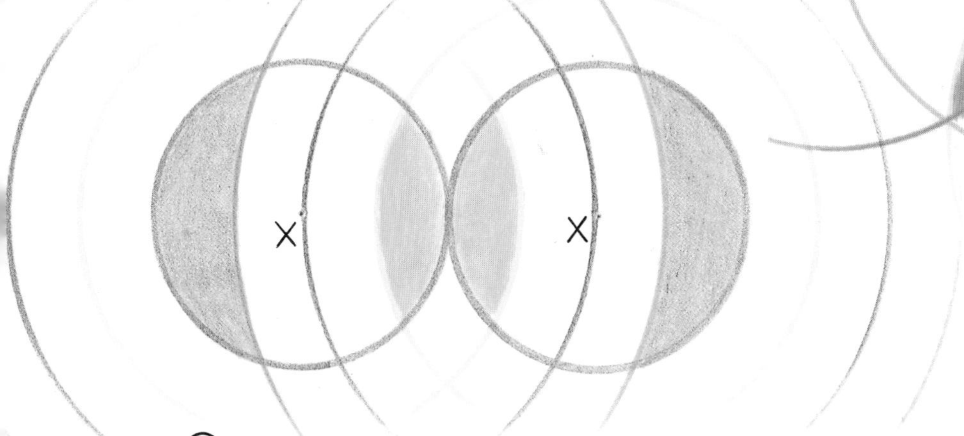

X X

② Mark two crosses X X, either 2 or 3 or 4 cm apart, near the middle of a piece of paper. Set your compasses at 2 cm, and draw a circle round each cross. Then set your compasses at 3 cm, and draw two more circles. Repeat at 4 cm, 5 cm, and so on. Fill in some of the overlapping shapes with different colours to make the patterns clearer.

What happens if you put your crosses closer together, or further apart? What happens if you start with three crosses in a line, with 3 cm gaps?

③ Draw a circle about 4 or 5 cm in diameter. Mark one point on the circle with an X. Now put the point of your compasses at a different place on the circle. Draw a new circle which goes through X. Repeat this several times, putting the point of your compasses in different places on the first circle, and drawing each new circle through X. What sort of shape do you get?

X

What happens if you start with a square, or some other shape, and then draw circles through one point on its perimeter?

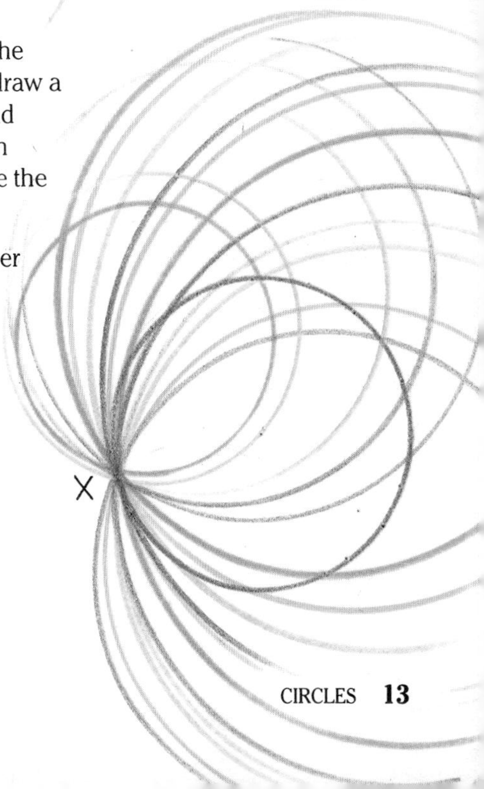

The balanced diet

Can you help a fat grocer with a weighty problem on Christmas Eve?

Twas the night before Christmas; he'd shut up the shop.
He'd had nothing to eat since his lunchtime lamb chop;
And all that was left of his produce and goods
Was a turkey, some onions, and three Christmas puds.

"My diet must be balanced!" he said to himself
As he took all the leftovers down from the shelf.
And the gleam of the scales inspired him to see
Just how balanced his diet could possibly be.

Three puds and the turkey he put on one tray;
On the other, 12 onions; the same they did weigh.
Next, on the left pan he put just the hen.
With three puds and six onions it balanced again.

It was then that he saw he was in a right fix,
Because onions and puds are a terrible mix.
So he left out the puds, and used onions alone.
How many will balance the bird on its own?

The problem is, how many onions does the turkey weigh? One way to solve it would be to go and buy a turkey, three Christmas puddings, and some onions . . . but that would be rather expensive.

You could guess. Suppose the answer is seven.

① Does that work? Go back through the first weighing and the second, and see if you can get the amounts to add up.

You could do it by reasoning. Both weighings have a turkey. The difference is that three puds are on the same side as the turkey in the first, and on the other in the second. The difference between these two weighings is six onions.

② So how many onions does a pudding weigh? And can you solve the rest of the problem?

This is fine, but hard to keep in your head. The easiest way to solve problems like this is to use equations. Equations allow you to write the problem down clearly, and tackle it in logical steps. So write down each weighing as an equation. The first looks like this:

1 Turkey + 3 Puds = 12 Onions

The middle of the equation is the equals sign. The left-hand side balances the right-hand side, just as it does on the kitchen scales. And the most important rule for equations is: ***keep the two sides balanced****. Whatever you do to one side, you must do the same to the other; that's the key to balancing an equation.*

You could add 2 more onions to each side.

1 Turkey + 3 Puds + 2 Onions = 14 Onions

It still balances. You don't need to do this with scales; you can see that it will work by looking at the equation.

Now let's look at both the weighings together:

$$\begin{array}{lll} & \textit{1 Turkey + 3 Puds} = \textit{12 Onions} \\ \textit{and} & \textit{1 Turkey} \qquad\qquad = \textit{6 Onions + 3 Puds} \end{array}$$

One thing you can do is add the two equations together. Add together everything in the two left pans, and it will still balance everything in the two right pans

2 Turkeys + 3 Puds = 18 Onions + 3 Puds

Now take 3 puddings out of each pan—or away from each side:

2 Turkeys = 18 Onions

(3) How many onions does a turkey weigh?

When you have two equations, you don't always have to add them. You may get your answer more quickly if you try something else. Try subtracting one from the other:

$$\begin{array}{rcl} \textit{1 Turkey + 3 Puds} & = & \textit{12 Onions} \\ -(\textit{1 Turkey} \qquad\qquad) & = & -(\textit{ 6 Onions + 3 Puds}) \\ \hline \textit{3 Puds} & = & \textit{6 Onions} - \textit{3 Puds} \end{array}$$

(4) Now try adding 3 Puds to each side of this equation. What do you get? Can you work out how many onions a Christmas pudding weighs? Does that tell you the weight of a turkey?

Try using equations to solve this problem:

(5) A brick balances a kilogram and half a brick. How many kilograms would balance one brick?

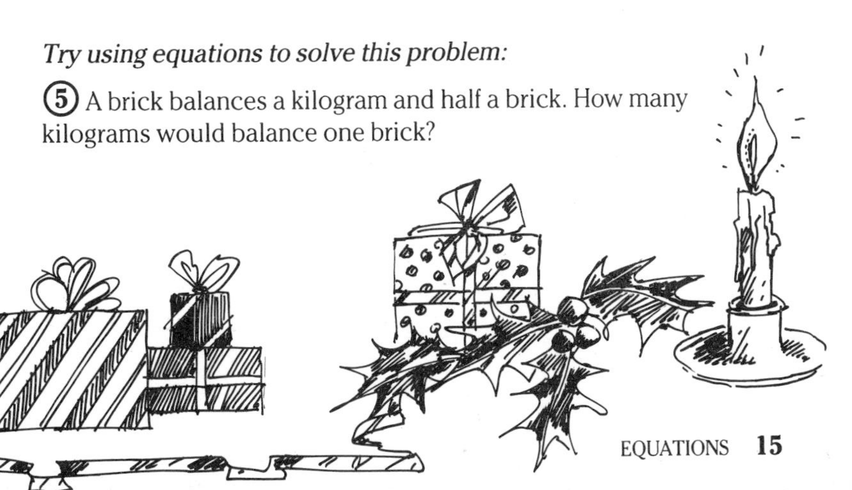

The Great Unknown

Twelve hundred years ago, a brilliant Arab mathematician called al-Khowarizmi used the word al-jabr to describe what he was doing. Today we call it algebra. The idea of algebra is to think about numbers we don't know instead of those we do. The Arabs used 'the Heap' for the number they did not know.

If I add 3 to the Heap, then I have 10.

What al-Khowarizmi meant was the Heap had 7 to start with.

① Take 6 from the Heap, and leave 16. How big was the Heap?

There are several ways to tackle these problems, including sensible trial and error, and logical reasoning.

You might realise that the Heap must be bigger than 16, since when you take away from it you are left with 16. So you might make a trial of 17 for the Heap. Now take away 6, and see what you are left with. Answer: 11—an error; so make another trial.

Think logically about the numbers. You might reason that if the new Heap is 6 smaller than the old one, then the old Heap must have been 6 bigger than the new one. This gives you the answer. However, when the problems are slightly harder, the reasoning becomes difficult to keep in your head.

② If 10 were added to the Heap it would be twice as big. How big is the Heap?

A third way to tackle these problems is by using equations. Equations allow you to write down things you know about numbers you don't know. For example, you can write question 2 as an equation:

$$10 + \text{Heap} = 2 \times \text{Heap}$$

This does not change the problem, but it may help you to look at it more clearly. You may be able to solve it at once.

Writing the equation gives you other chances too. You can push the equation around mechanically. You'll be glad to hear this sometimes lets you work out the answer without using any brainpower at all!

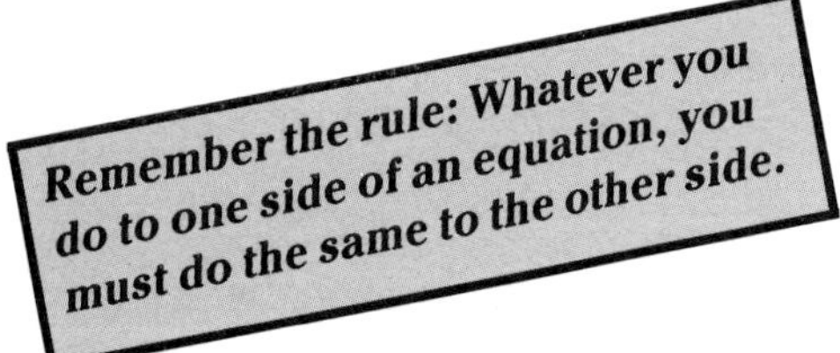

Remember the rule: Whatever you do to one side of an equation, you must do the same to the other side.

Take away Heap from each side of the equation above:

$$10 + \text{Heap} - \text{Heap} = (2 \times \text{Heap}) - \text{Heap}$$

Or $10 = \text{Heap}$

In other words, the Heap is equal to 10.

Today we most often use x *for the great unknown.* x *is like the Heap; it just stands for a number we don't know.*

③ If $x - 4 = 10$, what is x?

Again, you may be able to solve this in your head. If not, write down the equation and try adding 4 to each side.

Another advantage of algebra is that it allows you to talk about numbers that aren't fixed. Suppose James is five years younger than his sister Mary. Can we write down a rule for Mary's age?

We don't know how old James is. We can't give his age a number. But we can borrow al-Khowarizmi's idea and give his age a letter. Call James's age x. *Now we can talk about a number we don't know.*

Let James's age be x. *Then Mary's age must be 5 more than* x.

So Mary's age is $x + 5$.

④ Write down how old Mary will be when James is 20, 25, 55 and J years old.

⑤ Think of a number. Add 3. Take away 2. Take away the number you first thought of. Is the answer 1?

Will the answer always be 1? How can you tell? Try writing down the puzzle as an equation. Does that help you to see why it works?

⑥ Think of a number. Add 3. Multiply the result by 10. Take away 20. Divide by 5. Take away twice the number you first thought of. Is your answer 2?

Will it always be 2? Can you show why, by writing the puzzle as an equation?

⑦ Write your own 'Think of a number' puzzle, and use an equation to check that it will work.

Puzzle-cracking

① In ten years' time, Addie will be twice as old as she was five years ago. How old is she now?

② James has as many sisters as he has brothers. But his sister Jenny has twice as many brothers as she has sisters. How many children are there in the family?

Can you do these puzzles? There are lots and lots like them.

③ Two children have been collecting pennies, and have laid them out to see how long a line they make.

Farouq has half as many as Kate.
Together they have 48.

How many pence did each child collect?

You may find this puzzle easier to think about if you write it down as an equation. Call Farouq's number of pennies F. Kate has twice as many; so she has 2F.

Now write an equation for the total amount they have.

Use the equation to work out what F is. Write down how many each child has, and make sure the total is 48.

④ There were some spiders and some flies in a room, and the total number of legs was 48. (Spiders have eight legs; flies have six.) What numbers of each animal might there have been?

There are at least three possible answers. Write them all down in a table like this:

No of spiders	Spiders' legs	No of flies	Flies' legs	Total legs

Suppose each spider eats one fly, and then the total number of legs is 30. Now can you work out exactly how many spiders and how many flies there were to begin with?

⑤ Try solving the spiders and flies puzzle by logical argument. Copy and complete each of these sentences:

Number of legs eaten is $48 - 30 = \square$.

So number of flies eaten $= \square \div 6 = \triangle$.

But each spider ate one fly; so number of spiders is $\triangle$.

This number of spiders must have had $\triangle \times 8 = \triangleright$ legs

So out of the original 48 legs, $48 - \triangleright = \triangleleft$ must have been flies' legs.

So total number of flies to begin with must have been $\triangleleft \div 6 = \triangledown$.

So to begin with there were $\triangle$ spiders and $\triangledown$ flies.

⑥ Now try solving the same problem by using equations. Call the number of spiders S. Then you know that, to begin with, the total number of spiders' legs is 8S, and the total number of flies' legs is (48 − 8S).

You also know that when each spider eats one fly, they consume a total of . . . (how many legs?).

Go on from there, and see if you can work out the answer.

⑦ Phong has no sisters, but four brothers. The oldest is twice as old as the youngest. There is a two-year gap between each of the children, and Phong is in the middle. How old is she?

⑧ Here's a really difficult puzzle that can be solved by using an equation. The hands of a clock point in exactly the same direction at 12 noon. At five past one, they are nearly in a line again, but not quite, because the hour hand has moved on a little past the 1. What is the first time after noon when the hands again point in exactly the same direction?

Even the odd numbers

In the road where I live there are some 'ghost houses'. In fact it's 'haunted'—by all the missing numbers.

In many streets, the house numbers start 1, 3, 5 up the left-hand side, and 2, 4, 6 up the right.

① On which side would you expect to find number 11?

② Are the streets near you like this? Are all streets numbered in the same way?

Sometimes a few numbers are missing. These are the ghost houses. Either they haven't been built yet, or else they've fallen down. Then there's a gap in the numbers. They might go 2, . . . 6, 8, 10. But still the numbers come in an obvious order.

③ Why do you think it's useful to arrange house numbers like this?

④ Suppose you had a paper round. How would you arrange your newspapers in order to cross the road as few times as possible?

What number is your house? Are you on the right-hand side of the street, or on the left?

I live in number 35. That's an odd number.

1, 3, 5, 7 and so on are called the odd numbers. The even numbers are 2, 4, 6, 8, 10 and so on.

⑤ Make two columns in your book; one for odd and one for even numbers. Put each of these in the correct column: 7, 8, 78, 87, 98, 185, 999, 1000, 1001, 2223, 3332, 0. How do you know which goes in which column?

⑥ Just suppose you were to take a large piece of paper, and tear it into three pieces. Then you tear any one of those pieces into three more pieces. Suppose you can go on doing this as long as you like, but every time you tear a piece it must be into three.

Could you get a total of 10 pieces in this way? Or a total of 2000? What sorts of different totals are possible?

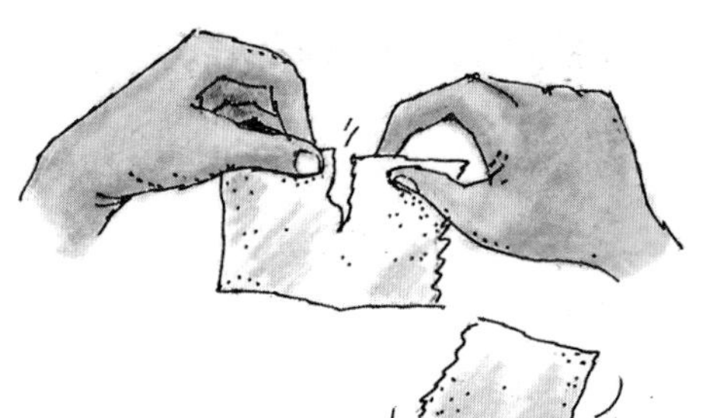

⑦ What happens when odd and even numbers are combined? Is the total always odd or always even when you

(a) add two odd numbers together,
(b) add an odd number to an even number,
(c) multiply odd by odd,
(d) multiply odd by even,
(e) take away one from the other?

Write your results in a table. Use symbols—perhaps O for any odd number and E for any even number, so that O + E means 'odd number plus even number'.

Are you sure that all the results are always true? Or are there some that you can't be certain about?

⑧ Suppose you sent off for a DIY coffee-table from a mail-order company. When it arrives the instructions are all in Mushipese, but you want to put it together. There is an obvious top, and some legs.

The legs are in seven straight pieces, which fit together end to end. The shortest piece is 10 cm long, the second is 20 cm, the third is 30 cm, and so on up to 70 cm. Can you sort out the legs, and build the table so that the top is level?

If you used only the first three pieces, you could make two legs of the same height (30 cm)—but that would be a bit low for a table. And would it stand up with only two legs?

Could you make two legs of the same height using all the first five pieces? What about the first six, or all seven?

Could you make three legs of equal height? Or four? Or five? Investigate how many different tables you could make, and write down which leg pieces you would use for each one.

Prime suspects

*A **prime** number has only two factors—itself and 1.*

Take the number 6. It has four factors: 1, 2, 3, and 6, because $1 \times 6 = 6$ and $2 \times 3 = 6$.

The number 4 has only three factors: 1, 2, and 4: $1 \times 4 = 4$, and $2 \times 2 = 4$.

① Copy this table and complete it as far as 20.

Number	Factors	Number of factors
1	1	1
2	1, 2	2
3	1, 3	2
4	1, 2, 4	3
5	1, 5	2
6	1, 2, 3, 6	4

Make a list of all the numbers that have only two factors. There should be eight of them below 20. What sort of numbers are they?

Many numbers can be shown as rectangles like this:

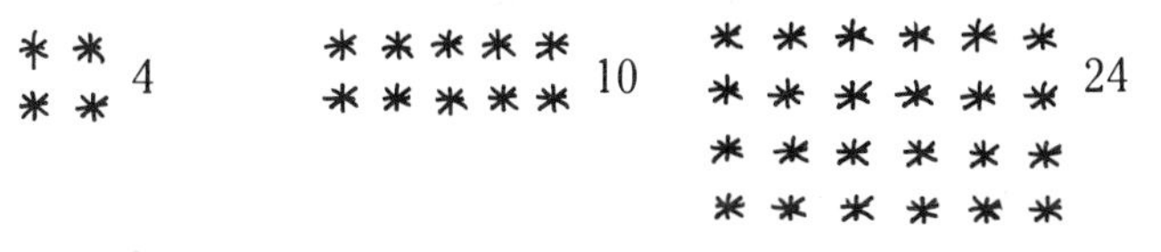

② Which numbers below 20 can't be shown as rectangles?

③ Separate the numbers below into two lists. Put the primes in one, and the rest, with their factors, in the other.

19, 29, 39, 49, 59, 69, 79, 89, 99, 199, 201.

④ From your table in question 1, make a list of the numbers that have an odd number of factors. Can you predict what the next two will be? Check them out to make sure.

What sort of rectangles do these numbers make?

⑤ The Number Police are worrying about 287. Write a note to explain to them how to find out whether a number is prime or not.

⑥ Multiply the first two prime numbers together and add 1. That is $(2 \times 3) = 6$, and $(6 + 1) = 7$. Because we added 1, the answer, 7, can't be exactly divisible by either 2 or 3.

Now suppose you use more primes: $(2 \times 3 \times 5 \times 7) + 1$. The answer can't be divisible by 2 or 3 or 5 or 7.

What would happen if you went on doing this? Would you get bigger and bigger primes?

Do you think there could be a biggest prime number?

Fibonacci and his rabbits

One of the first really brilliant mathematicians of western Europe lived in Pisa about AD 1200—while the leaning tower was being built. His name was Leonardo, but his dad's name was Bonaccio, and they called him Filius Bonacci, or Fibonacci for short.

In his first maths book, published in 1202, Fibonacci wrote a problem about rabbits.

Suppose a pair of baby rabbits grow up in one month, and produce a pair of babies when they are two months old. Then they produce another pair every month. Meanwhile the second pair grow up, and start producing their own babies, and so on.

Start with one pair of babies on 1 January. How many pairs of rabbits will you have on the first day of each month?

① See if you can work it out before you read the next bit.

On 1 January there was only 1 pair. Let's call them **a** *&* **b**.

By 1 February they will have grown up to adults **A** *&* **B**, *but they won't have had any babies; so there will still be only 1 pair.*

By 1 March **A** *&* **B** *will have had babies* **c** *&* **d**; *so there will be two pairs altogether.*

By 1 April **A** *&* **B** *will have produced* **e** *&* **f**; *so there will be three pairs—***A** *&* **B**, **C** *&* **D** *(now adults), and* **e** *&* **f**.

By 1 May the two oldest pairs, **A** *&* **B** *and* **C** *&* **D**, *will both have had babies; so there will be five pairs:* **A** *&* **B**, **C** *&* **D**, **E** *&* **F**, **g** *&* **h**, *and* **i** *&* **j**.

By 1 June the first three adult pairs will have produced three new pairs of babies, making eight pairs altogether.

When you work it all out, the Fibonacci sequence begins

1 1 2 3 5 8 13 21 34 55 89 144 ...

This number sequence is as full of patterns as Fibonacci's house must have been full of rabbits.

② Try adding together any pair of next-door numbers in the Fibonacci sequence. What do you get? Try another pair, and a third. What do you notice about the answers? The sequence above gives the first twelve Fibonacci numbers. Can you predict the next four?

③ Try adding together the 1st and the 3rd Fibonacci numbers. Then the 1st and the 3rd and the 5th. What do you notice?

④ Pineapples and pine cones have their segments arranged in spirals; some go to the left, the others go to the right. In the photograph two spirals are highlighted, one in each direction.

At the top two opposite spirals begin together. How many segments do you have to count down each until they cross again? Where have you seen these numbers before?

Find out in the library where else Fibonacci numbers turn up in Nature.

⑤ The table below shows the results of dividing each Fibonacci number by the one before it in the sequence. Copy and complete the table, using a calculator. Where possible, write your answers rounded to three decimal places.

1	1	2	3	5	8	13	21	34	55	89
	1	1	2	3	5	8	13	21	34	55
	1	2	1·5	1·667	1·6	?	?	?	?	?

As you go further along the sequence, the answers get closer and closer to 1·618. This is called the Golden Ratio, or the Golden Section. It turns up a lot in art and architecture. See what you can find out about it in the library.

⑥ Take a big piece of squared paper. Draw on it the biggest rectangle you can, using Fibonacci numbers for the length of each side. For example, it might be 55 by 34 squares.

Complete a 34 by 34 square, which will give you another Fibonacci rectangle, 34 by 21.

Now complete a 21 by 21 square to get another Fibonacci rectangle, and so on. You should be able to fill the whole space with Fibonacci rectangles. That's how this knitting pattern was made.

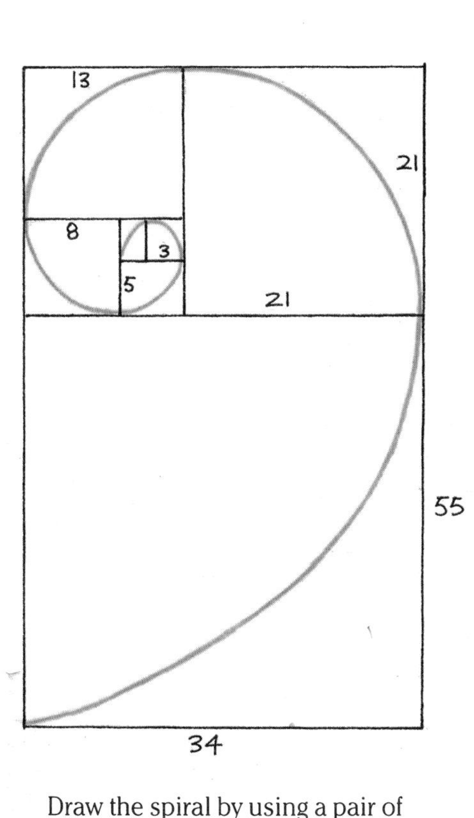

Draw the spiral by using a pair of compasses to draw a quarter-circle in each square, starting at the outside.

⑦ See what other patterns you can find in the Fibonacci sequence.

Half-witted

Three fat friends were famished; so they bought a big pork pie,
And took it off to cut up into lumps.
But how could they divide it so that none of them would cry,
And no one would be left down in the dumps?

What the fat friends wanted was proper pie-division;
So each of them should have an equal slice.
They cut it down the middle—with a penknife and precision—
And then the same again, since that looked nice.

But they didn't get three pieces, and they couldn't work out how
To make three slices all of equal size.
And just as their discussion had become a mighty row,
The dog scoffed all four bits before their eyes!

The fat friends should have been happy. They had a big pork pie. But they lacked one vital ingredient. They wanted to be fair, and to be able to share the pie out evenly. They were unhappy, because they didn't understand ***fractions****.*

A fraction means a bit of something, but not the whole thing. The fraction is sometimes more than the whole thing; people in the pub order 'three halves of bitter'. More often, a fraction means only a part. A mouthful of Coke is a fraction of the can. A few crisps is a fraction of the packet. The tail is a fraction of the cat.

Cut the pie in two equal pieces, and you get two halves. A half is a fraction. We write one half as $\frac{1}{2}$*, because we have divided 1 pie into 2 pieces.*

① Suppose each half of the pie is cut into two equal pieces. How many pieces will there be altogether? What fraction is each piece?

② Americans have a 25-cent coin. Why do you think they call it a 'quarter'?

Imagine that you had a pie. Before you cut it, think! You can divide it into as many equal shares as you want. How big is your gang?

③ Copy and complete this sentence: 1 pie = 2 halves, 3 ______, or __ quarters, or 5 ______, or __ sixths, or 7 ______, or __ eighths, or 9 ______, or __ tenths.

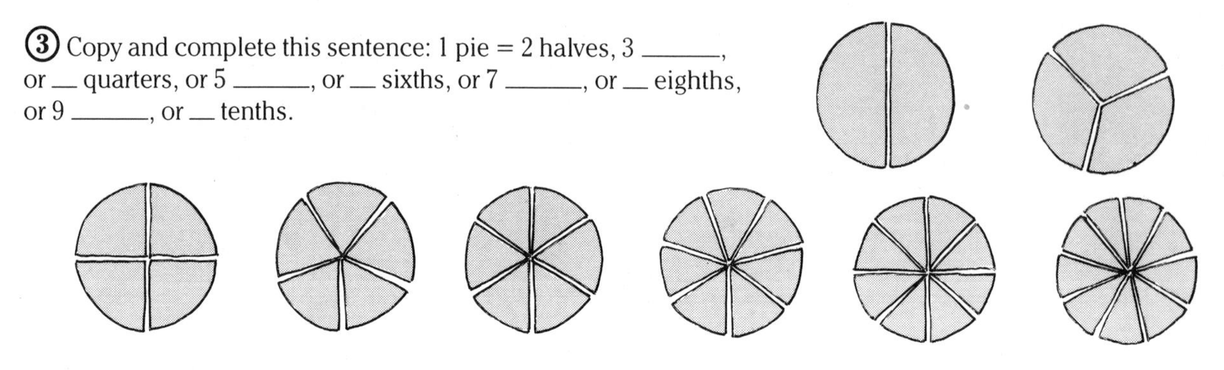

In a fraction, the number below the line is the number of slices. When a pie is cut into three equal pieces, each piece is $\frac{1}{3}$ of the pie.

The bigger the number below the line, the smaller the slice. This is because the more slices you cut, the smaller each one has to be. So $\frac{1}{5}$ of a pie is smaller than $\frac{1}{3}$ of the same pie.

④ Write these fractions in order of increasing size: $\frac{1}{3}, \frac{1}{10}, \frac{1}{2}, \frac{1}{8}, \frac{1}{4}$. Explain why the last one in your list is the biggest. Draw a pie, and then, for each fraction, draw a slice of roughly the right size.

Take more than one slice, and you get a bigger fraction. Suppose a pie is divided into three, and a greedy-guts takes two slices; that is two-thirds of the pie. We write this as $\frac{2}{3}$.

When two fractions have the same number below the line—$\frac{3}{5}$ and $\frac{4}{5}$, for example—then the bigger the number on top, the bigger the fraction. So $\frac{4}{5}$ is more than $\frac{3}{5}$, which is more than $\frac{2}{5}$, and so on.

⑤ Write down the following fractions in order of increasing size: $\frac{3}{4}, \frac{3}{5}, \frac{1}{4}, \frac{1}{2}, \frac{99}{100}, \frac{9}{10}, \frac{1}{17}$. Try doing this first without a calculator. Then do it with a calculator, and see whether you get the same order.

⑥ The three fat friends cut the pie into four equal slices. If the greediest had taken two of the slices, what fraction of the pie would he have had? What fraction would each of the others have had?

⑦ If you divide a pie into 100 slices, what fraction is each slice? What per cent of the pie is each slice? What per cent of the pie *should* each of the three friends have taken?

⑧ Suppose the three friends had put their money together to buy the pie, and had been given £1·17 change. How should they have divided the change equally? What is one third of £1·17?

⑨ Fifty per cent of a cake is exactly $\frac{1}{2}$. Ten per cent is $\frac{1}{10}$. Find out what other percentages are simple fractions. Draw up a table to show the relationships between percentages and fractions.

Numbers with a point

In the remote town of Tensonly they don't like fractions. There are no percentages. Everything happens in tens. It's a strictly decimal town. That's why they call it Tensonly.

The only fractions allowed in Tensonly are tenths, hundredths (tenths of tenths), and so on. Even then, they don't write them as fractions, but as decimal numbers.

The decimal point is a signal to show where the whole part of the number ends. Everything to the right of the decimal point is worth less than one. The first digit after the point is the number of tenths; the first two are the number of hundredths, and so on.

So they write 3·4 when they mean three and four tenths, and 3·42 when they mean three and forty-two hundredths.

① Write $1\frac{4}{10}$ and $12\frac{6}{10}$ as decimal numbers.

To turn other fractions into decimals, the people of Tensonly use calculators to divide the number above the line by the number underneath. So $1\frac{2}{5}$ *is* $1 + \frac{2}{5}$*, or 1·4.*

② Write down the decimal versions of $2\frac{1}{2}$ and $5\frac{1}{5}$.

In a decimal number, each move to the right of the decimal point divides the digit by an extra 10. So 4·25 means $4 + \frac{2}{10} + \frac{5}{10 \times 10}$ *(four units, two tenths, and five hundredths).*

Money presents no problems in Tensonly, because it's already written in decimal form. If a cassette tape is £2·49, then its cost in pounds is $2 \times 1 + \frac{4}{10} + \frac{9}{10 \times 10}$.

£1 is 100 pence. To work out a cost in pence all you have to do is multiply the cost in pounds by 100. This moves all the digits two places to the left, because all the hundredths become whole numbers.

Cost in pence: 100 × £2·49 = 249p

③ What is the cost in pounds of one hundred 22p stamps? What about fifty 24p stamps?

④ Write down what happens when you divide a decimal number by ten. How many days are there in one tenth of a year? How many days in one tenth of a week?

Time is a bit of a problem in Tensonly. We have 24 hours in a day, and 60 minutes in an hour, but they use only tens.

5 Design a suitable clock for use in Tensonly kitchens. Write ten lines to explain its special features.

6 The only real game in Tensonly is tennis! How would you score at tennis, using only tens?

One problem about decimal numbers is that some of them are endless. For example, divide 1 by 3. The answer is not 0·3, nor 0·33, nor 0·3333, nor even 0·33333333333333333333333333333333333333. The answer is one third, but you cannot write $\frac{1}{3}$ exactly in simple decimal form, because the 3's go on for ever.

7 Use your calculator to divide 1 by 3. How many of those 3's do you think you really need? Write a short form of the decimal that would be accurate enough in real life. Now do the same for $\frac{2}{3}$.

8 A recurring decimal has an endless repeating pattern—like the 3's above. Try dividing 1 by each number from 1 to 20. Write a table of what you get. Which ones are recurring decimals?

9 Use a calculator to find the decimal form of these fractions: $\frac{1}{7}$, $\frac{2}{7}$, $\frac{3}{7}$, $\frac{4}{7}$, $\frac{5}{7}$, and $\frac{6}{7}$. Write them down in a table. What patterns can you see?

10 How many different decimal numbers can you write down that are larger than 1·2 but smaller than 1·4?

0.5 of a loaf is better than none.

The decimal clinic

Some people who are new to Tensonly can't cope with decimals. They get dreadful decimal diseases, and dire decimal distress. So there is a decimal clinic, to help people understand that numbers aren't pointless.

The most common problem is Decimal Blindness.

① Write down the right answer to 1·2 + 1·3.

② Explain what Decimal Blindness means, and why it led to a mistake. Suggest another sum that this patient might get wrong.

Other people lose their sense of place value. **Remember:** *the first digit after the decimal point shows tenths; the second shows hundredths; the third shows thousandths; and so on. If you remember that, you can't fall for this one:*

0.2	0.4	0.6	0.8	0.10	0.12

③ Can you see what's wrong with this sequence? If you know, write it down. If you aren't sure, try putting 0·2 into your calculator; then add 0·2 again and again.

The fifth number should be 1·0, not 0·10, that is, ten tenths, not one tenth. When you have ten tenths you have one unit. So you put 1 in the units column, before the decimal point. Then you have nothing left on the right of the decimal point.

④ Here's another sequence: 0·3, 0·6, 0·9, ... Write down the next three numbers in the sequence. Try it on your calculator. Try it on your friends.

⑤ Write down these three numbers with the smallest first and the biggest last:

1·7   1·55   1·689

(Hint: look at the first digit after the decimal point.)

When you are sure you have the right answer, try adding 0·001 to each number, and see whether they are still in the same order.

Write three lines to explain why the 9 in 1·689 is the least important digit in these three numbers.

One more thing to avoid is the Dreaded Decimal Split. Some people fall into the trap of thinking a decimal number is not one number, but two separate numbers, with a point between them:

⑥ Write three lines to explain how these numbers should be added. The decimal point doesn't cut a number in two; so explain exactly what it does.

Graphic picture stories

Portraits are usually pictures of people; they tell you what faces look like. Most maps are pictures of land; they tell you what landmarks you can find if you visit the area.

Many graphs are pictures of events; they tell you what has happened and in what order. The one below tells the story of a steeplechase, with horses racing over jumps.

The 3.30 Graphic Stakes at Disaster was run by four horses: Knock-knees, Bumble, Tumble and Crash.

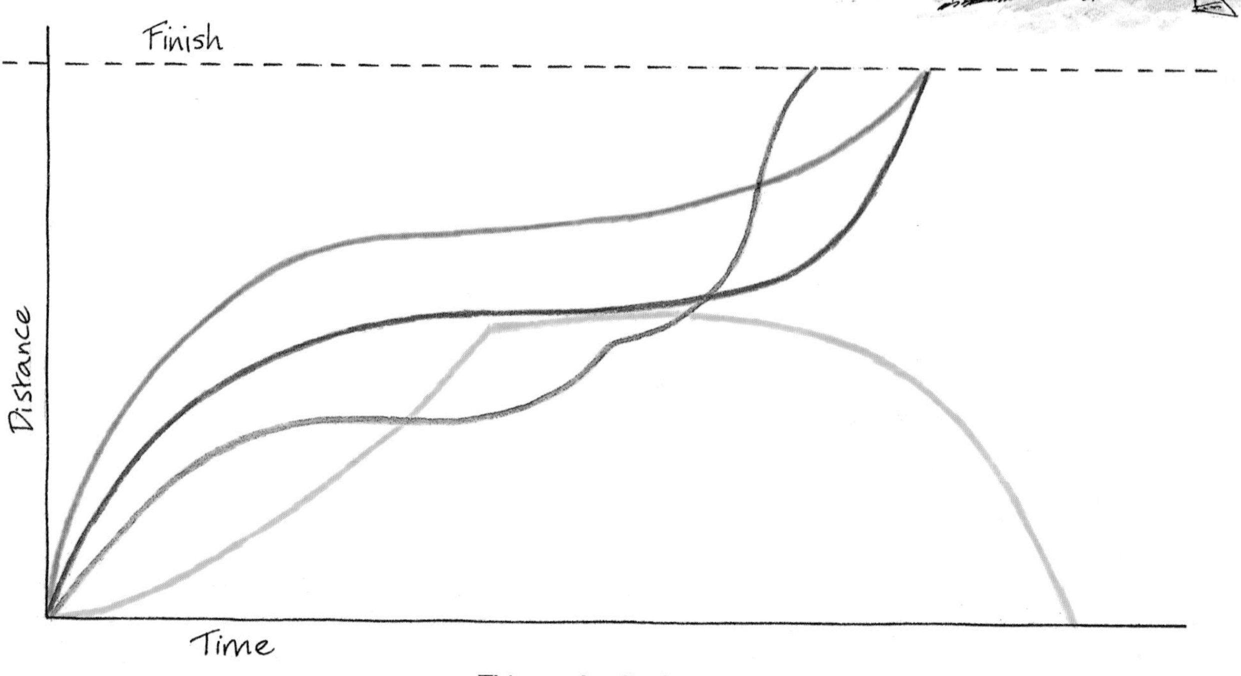

KNOCK-KNEES
BUMBLE
TUMBLE
CRASH

This graph tells the story of the race. Look at it carefully to work out what happened.

(Hint: choose a time, a little way into the race, and look straight up the page to find out what order the horses are in — which one has travelled the greatest distance? Then choose another time, a little bit later, and see what has changed.)

① Who wins the race? How do you know?

② Do all the horses reach the finishing line?

③ What do you think happens in the middle of the race?

④ Imagine that you are a radio commentator, broadcasting from Disaster. Write 20 lines of commentary about the race. Make it exciting, but don't forget to mention everything that happens. Remember, your listeners can't see the racecourse or the graph.

This graph tells the story of part of a game of tennis at Wimbledon.

The red line represents the speed of the tennis ball. The server is Steffi Graph, and the first thing she does is throw the ball up in the air. It's in her hand during the time AB, and it's getting faster all the time, until she lets go of it at the moment B.

Above her head the ball slows, because gravity is trying to pull it down again. So although the ball is still going up, its speed is getting lower and lower, as shown by BC. Remember that the upwards direction is speed, not distance.

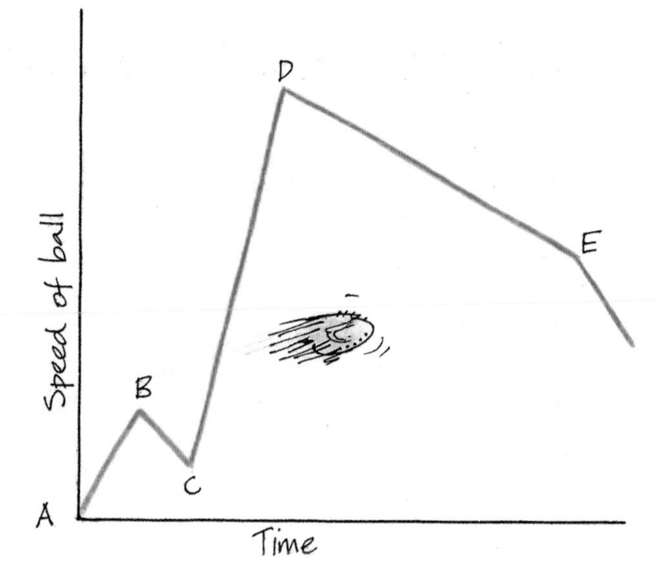

At the moment C, just as the ball reaches its highest point, she hits it with her racket. The racket stays in contact with the ball for the fraction of a second from C to D. All this time the racket is greatly increasing the speed of the ball. So the ball is moving much faster by the time it leaves the racket at D.

⑤ What happens at E, when the ball first hits the ground? (Hint: would you expect the ball to get slower or faster when it hits the ground? Which way does the red line go – up or down?)

Now your job is to make up the graph for the rest of the point.

⑥ Suppose the service is an ace, and the other player misses it completely. The ball bounces once more and is then stopped by the fence at the back. How could you show this on the graph?

⑦ Suppose the other player does return the serve, in a high lob. When will the speed of the ball be lowest? High in the air, or as it hits the ground? Sketch the graph of speed against time.

⑧ Make up your own story for the next point, and draw a graph of the speed of the ball. How would your graph show the server bouncing the ball on the ground before serving?

The next world record

Athletes keep breaking world records.

This is how some women's world-record speeds have changed over the last hundred years. These were sprints over 100 yards or 100 metres.

Year	1910	1921	1939	1950	1970	1982
Speed (m/s)	7·6	7·8	8·3	8·5	9·0	9·2

① What does this table tell you about how records are changing?

② Guess what the world record speed for 100 metres might have been in 1900, and what it might be in 2000.

The numbers above are called 'raw data'. Often you can see patterns more easily if you make a picture from the numbers. One way is to draw a graph.

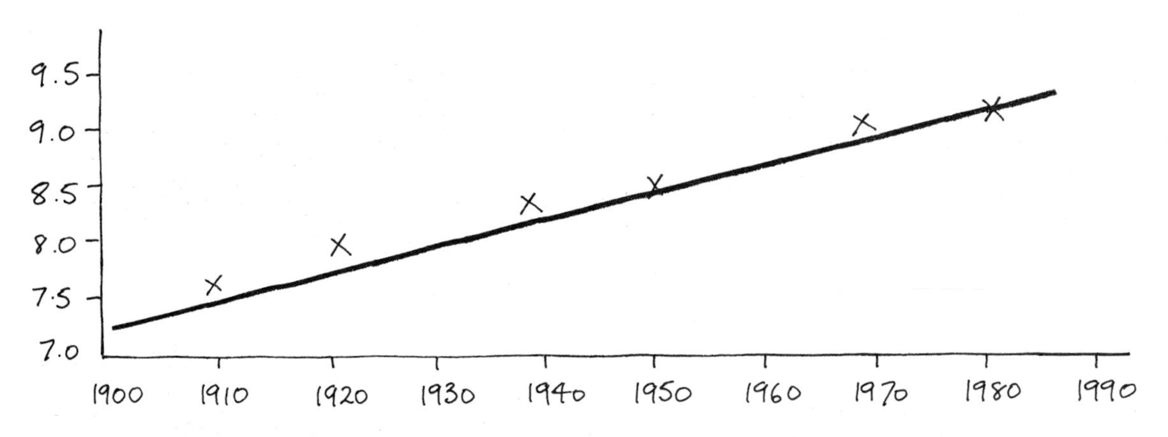

The graph shows you a picture of what is happening. It tells you the story of the world records over this distance.

③ The graph is almost a straight line. What does that tell you about the rate of change? Do the athletes show an even more dramatic improvement every ten years, or has their speed increased by about the same amount in each decade?

You can use the graph to estimate what the 100 metre speed might have been in 1900: the answer is about 7·2 m/s.

④ Estimate from the graph the value of the speed for the year 2000. Is it the same as the speed you guessed in question 2?

⑤ Suggest three possible reasons why world-record speeds in athletics are increasing. Do you think this will go on for ever?

Here are some world-record results for 1500 metres

	Women				Men				
Year	1922	1952	1962	1980	1900	1923	1941	1960	1980
Speed (m/s)	4·2	5·4	5·8	6·5	6·1	6·4	6·6	6·9	7·1

⑥ On one piece of squared paper, graph one line for the women's speeds and another for the men's speeds.

Write five lines to explain what the graphs tell you, and then answer these questions. Are women getting faster at the same rate as men? Do you think they will catch the men up? If so, when?

⑦ Use your graph to predict what the world-record speeds for 1500 metres might be at the next Olympic Games. Then work out what the world-record **times** would be.

What's happened?

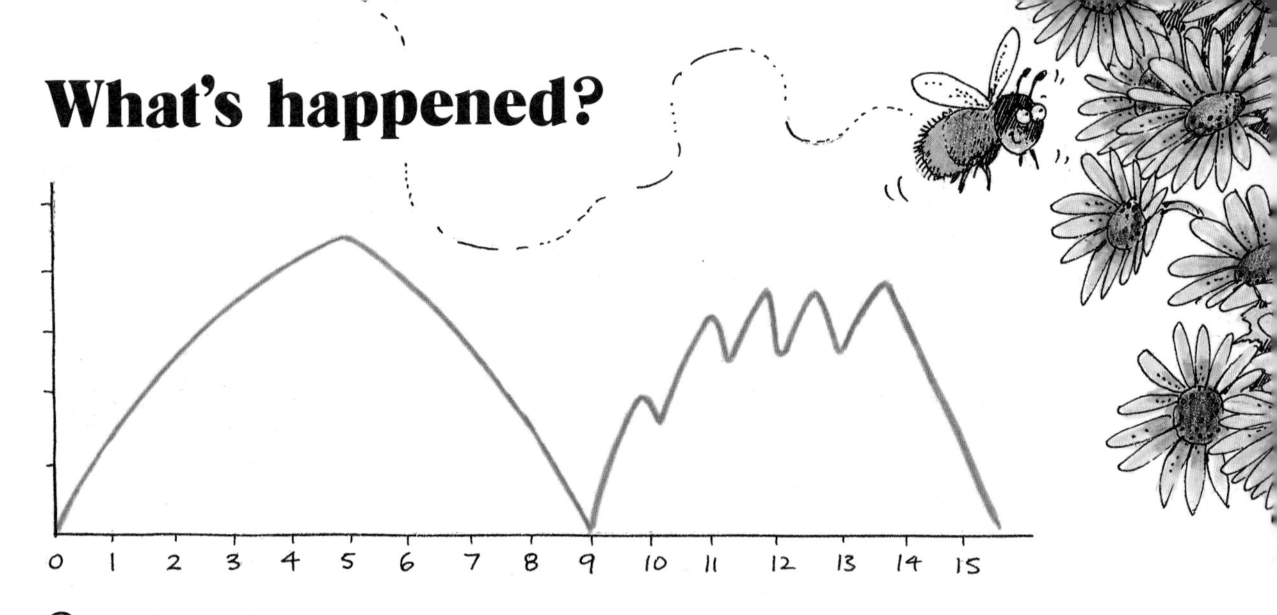

① A bee sets out from its hive to search for flowers that might have take-away nectar. This picture shows the distance the bee has travelled from the hive, and the time it has been away. Describe the bee's journey. What is it doing? What do you think the units might be for distance and for time?

② Could the same graph show how hungry a girl feels during school? Explain what the numbers might mean, and try to tell the girl's story. When would her tummy rumble most?

③ Amanda drives to work every day. The journey takes about a quarter of an hour. First she goes down the motorway. Then at the end of the motorway there's usually a traffic jam at the roundabout, and a series of narrow roads on her way to the office. Explain how the graph tells the story of her journey.

④ Bert fancies himself as a bathroom baritone. He can often hit the first high note clearly, but the second gives him problems. If the graph represents Bert's voice, what does it sound like?

⑤ The graph shows the number of people in an office who visit the coffee machine during the day. What can you guess about the way they work in the office?

⑥ Could this graph show the speed of swing of a pendulum? If not, what would that look like? At which point on a pendulum's swing is it moving most quickly?

You turn on both taps to run a bath. After 2 minutes the water is 12 cm deep. After 3 minutes it is 17 cm deep. After 5 minutes it is 27 cm deep. Then the telephone rings; it's your best friend, and you forget all about the bath while you chat.

⑦ (a) When should you have turned the taps off, if you like your bathwater 32 cm deep?

(b) How deep is the bathtub, if the water overflows after eight minutes?

⑧ The sketches below show the movements in four sports events. Match each picture with one of the following sports: 800 metres; javelin; 200 metres swimming; squash. Explain your reasons.

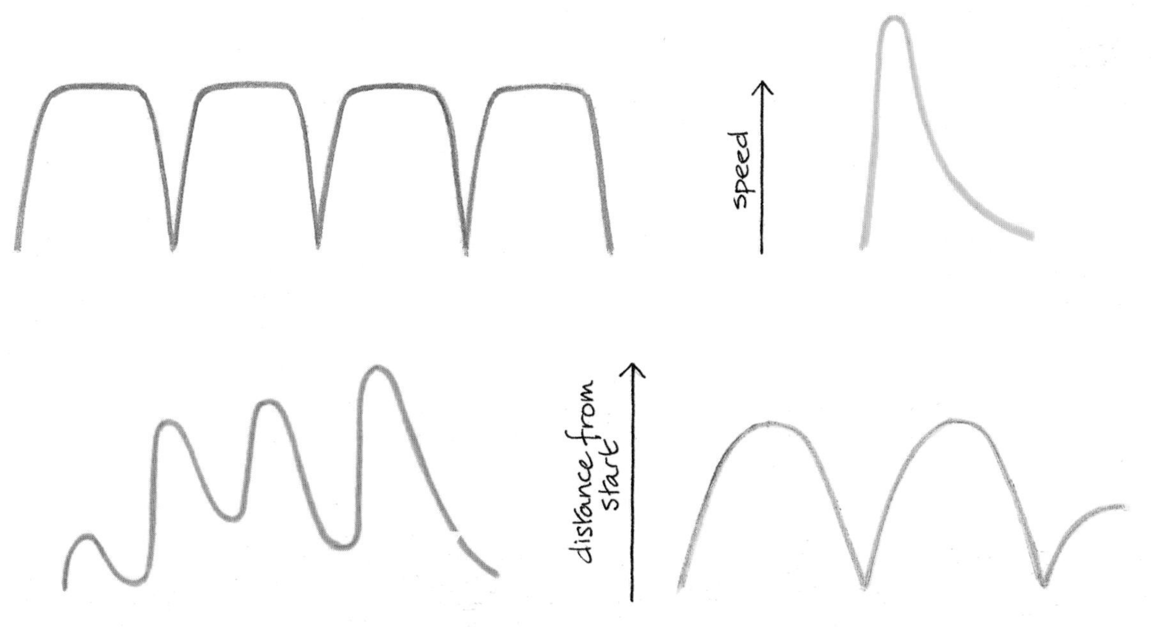

First steps in line-taming

Lines aren't always easy to handle. They can be difficult. They can be dangerous. They got Mr Pythagoras into trouble (see page 91). Be warned! Better still, be armed! Take a lesson in line-taming.

Suppose you want to draw on a flat piece of paper a clear picture of a solid box; you need skill. Or you can follow the steps below.

Step one—JOINING TWO SQUARES

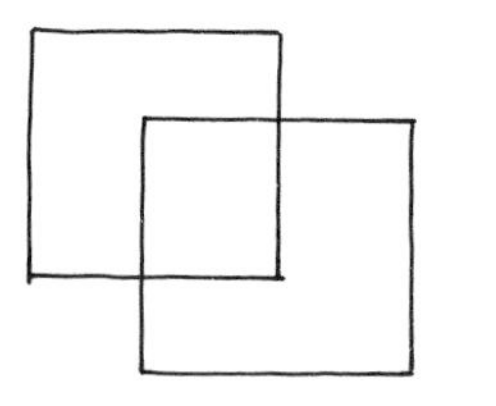

A *Draw two squares*

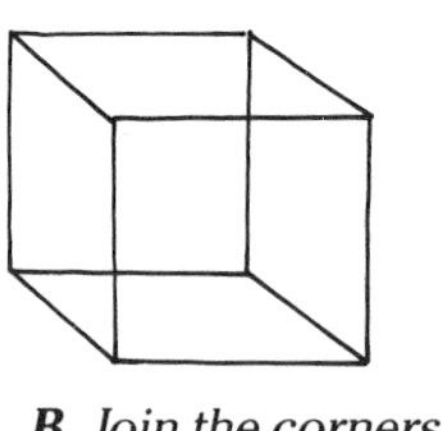

B *Join the corners*

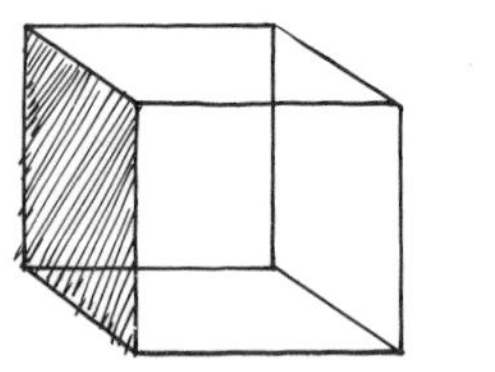

C *Shade a side*

D

D *Do this again, but leave out (or rub out) the 'hidden' lines that you wouldn't see if it was a real cube.*

This is called a vanishing point

Step two—ONE-POINT PERSPECTIVE

E *Draw a square*

F

G

I

F *Pick a point in the top corner of the page, and join it faintly to the three nearest corners of the square.*

G *Imagine the lines you have drawn chopped off a little way back from the square.*

H *Rub out the lines behind the chop.*

H

I *Join the loose ends, make the faint lines thicker to form edges, and shade to taste.*

Artists began to learn tricks like this about 500 years ago. Before then they realized that you couldn't see the backs of objects, and that the fronts looked more or less flat, but they had not worked out rules to make the sides look reasonable.

Vanishing points help artists to make things look right.

When you use simple one-point perspective, the front of the box must face you square on. Two-point perspective gets over that problem, and helps you draw an even more realistic picture.

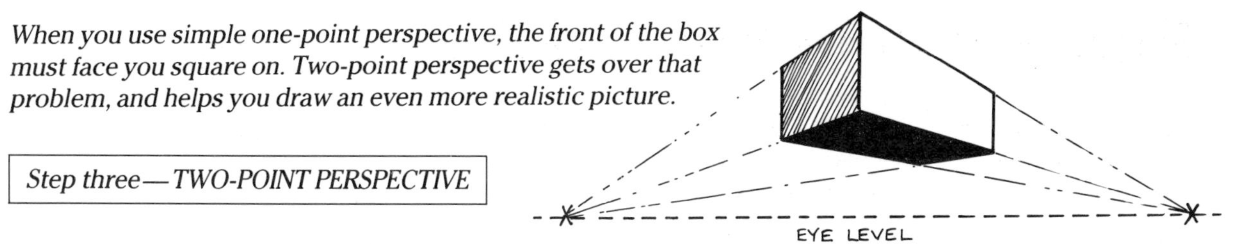

Step three— TWO-POINT PERSPECTIVE

J *First, pick two vanishing points.*

The line between them corresponds to your eye level.

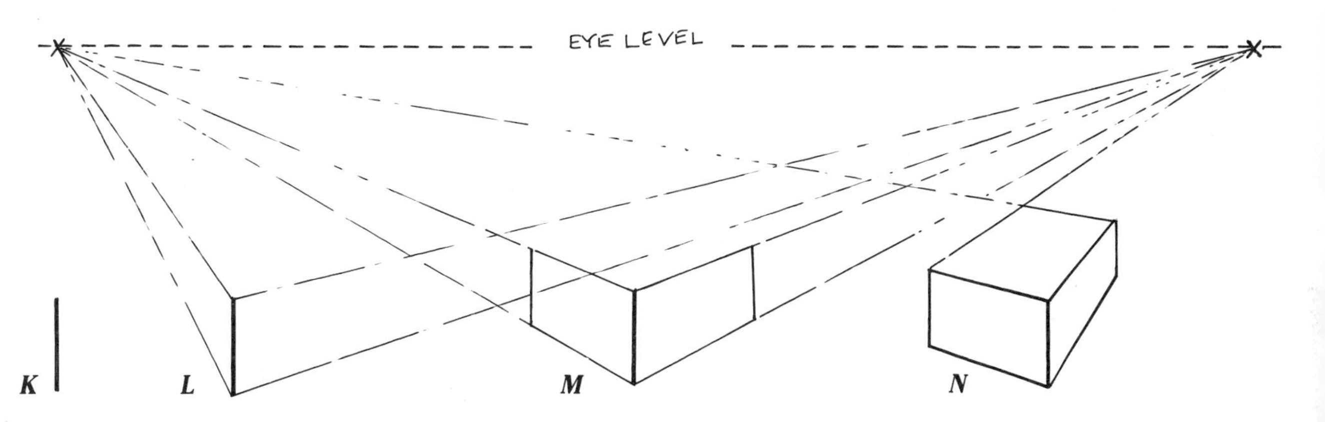

K *Draw a vertical line. This will be the box edge that is nearest to you.*

L *Very faintly, join both ends of the line to both vanishing points.*

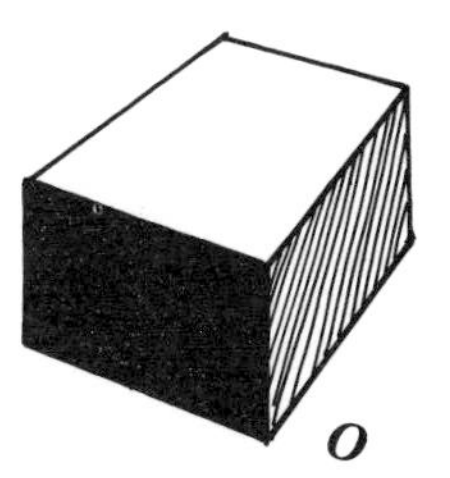

M *Chop off both pairs of lines (as in G) with new vertical lines.*

N *Very faintly, join the top of each of these new lines to the vanishing point on the opposite side.*

O *Draw over the edges of your box. Rub out the faint lines. Shade a side or two.*

① Draw a box by joining squares, a box by one-point perspective, and a box by two-point perspective. Which method do you like best? Which box do you like best? Why?

② Try drawing boxes of different sizes. Can you draw a small box inside a larger one?

③ Can you use these methods to draw buildings? Or cars? What else could you draw this way?

④ Using two-point perspective, what happens if you use vanishing points half way down the page? Or at the bottom? What if they aren't on the same level?

Barriers and lines

We use lines on paper to represent things in the real world. Some of these lines show imaginary things, but others represent solid barriers. If this line was a real barrier, the words would have to tunnel underneath to escape.

① Take some ordinary pencils. How many can you lay on the table so that each one touches all the others?

Laying them side by side is no good. Here the green one doesn't touch the yellow one.

You can put all the points together. Then you can get three touching like this, but not more than three.

Suppose you have two blue pencils touching at the points. You can bring up the red one to touch both. But if you try to bring in the yellow it can't touch the red, because the blue ones make a barrier.

The red and the yellow can touch points only if you pull the blue pencils apart first.

With pencils of different lengths, it's possible to get four all touching, but not at the points.

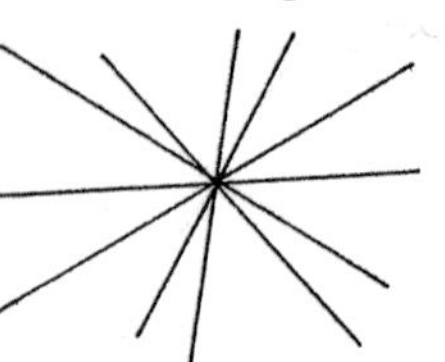

Lines are different. Here are six lines meeting at a point. You could draw six hundred to meet at the same point if you wanted. That's because lines don't make real barriers. We imagine that lines have no mass, and no thickness. Pencils make a real barrier across the table, because they have mass and thickness.

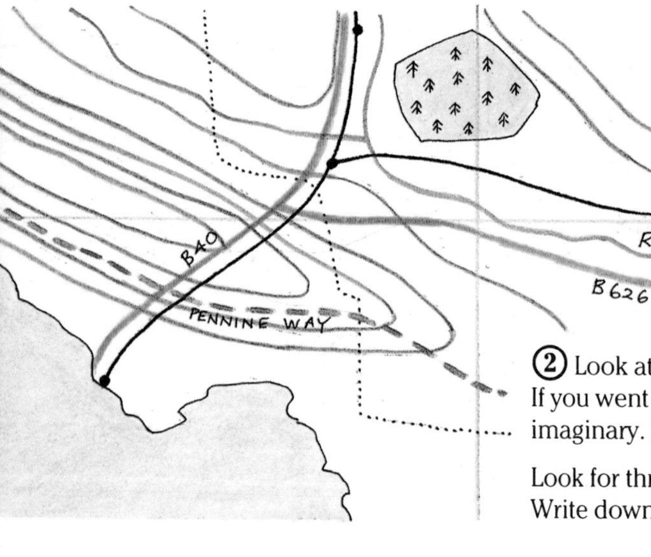

② Look at this map. Some of the lines on it represent real things. If you went to the actual places you would find them. Others are imaginary. If you went to look for them you wouldn't find anything.

Look for three 'real' lines and three 'imaginary' lines on the map. Write down what each imaginary line means.

The great experts on lines were the ancient Greeks. They used lines to survey and measure land. They called the earth 'geos' and measuring 'metros'; so they called land-measuring 'geometry'. Geometry is all about lines and how they work.

The geometry worked out by the ancient Greeks was all collected and written down on papyrus by a man called Eucleides, or Euclid for short. He was the first Professor of Maths in Alexandria. The 13 books of his Elements of Geometry *appeared about 300 BC, but are so clear and precise that we still use Euclid's geometry today.*

Surprisingly, Euclid was not very clear about straight lines. He said a straight line is 'a line that lies evenly with the points on itself.'

③ Suppose you were stranded on a desert island, with no ruler or straight edges. How could you draw a straight line in the sand? (Hint: Could you find a piece of string or thread somewhere?)

④ And how could you find whether three palm trees were in a straight line, even if one was on a different island? (Hint: this would be difficult in the dark!)

⑤ How long is the longest straight line that could be drawn on this page? (Hint: there are two places to draw the longest line.)

⑥ Write down clearly what you think a straight line is.

⑦ Imagine that these lines are real fences. There appear to be seven sheep in the field, but one is a wolf in sheep's clothing and may attack the others.

Where can you put just three more straight fences so that each animal is in a pen by itself?

⑧ Could you plant nine trees so that there are ten straight lines with three trees in each line?

Puzzling paper-round

You have just started a new paper-round. You collect a bundle of newspapers from Mr Patel, and you have to deliver them to houses in Bean Street, Dean Street, Jean Street, Kean Street, Lean Street, Mean Street, and Queen's Road. Each has houses on one side only.

You want to get round and home again before breakfast; so you must not waste time. You certainly don't want to go down any of these streets more than once, if you can avoid it.

(Don't mark this book. Trace or copy the map on to another piece of paper.)

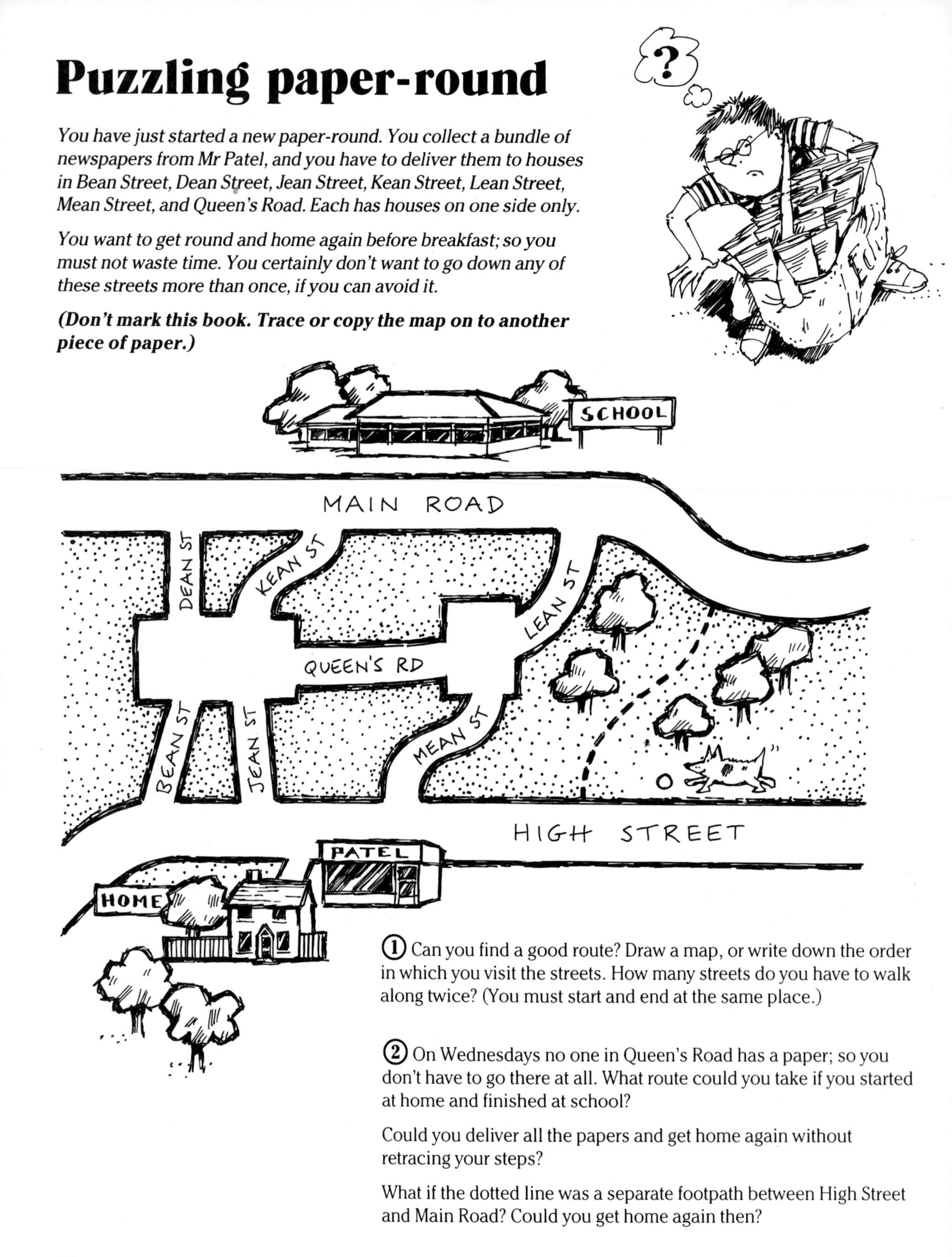

① Can you find a good route? Draw a map, or write down the order in which you visit the streets. How many streets do you have to walk along twice? (You must start and end at the same place.)

② On Wednesdays no one in Queen's Road has a paper; so you don't have to go there at all. What route could you take if you started at home and finished at school?

Could you deliver all the papers and get home again without retracing your steps?

What if the dotted line was a separate footpath between High Street and Main Road? Could you get home again then?

*This is a **network** problem. Finding a way through depends on how the roads join. The roads don't have to be straight. Their length doesn't matter. Their names don't matter. The only things that matter are the connections.*

One way to start solving the paper-round problem is to try a simpler problem first. Suppose there were just three avenues between West Park and East Park. You could easily go down each avenue just once.

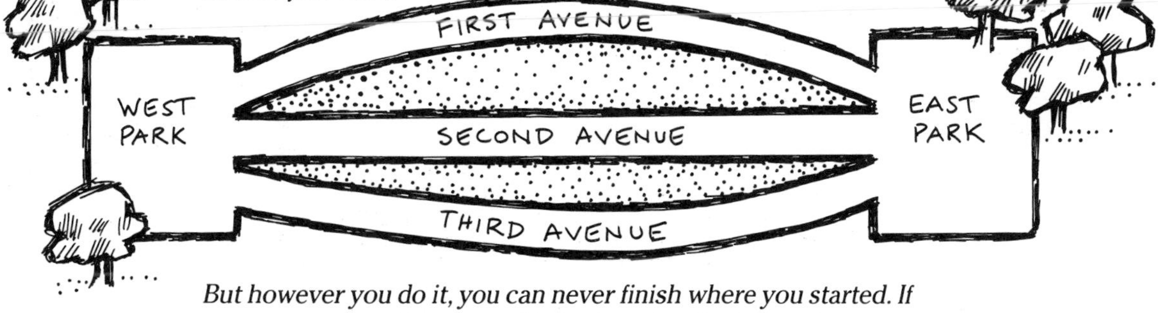

But however you do it, you can never finish where you started. If you start in West Park you will finish in East Park, and vice versa.

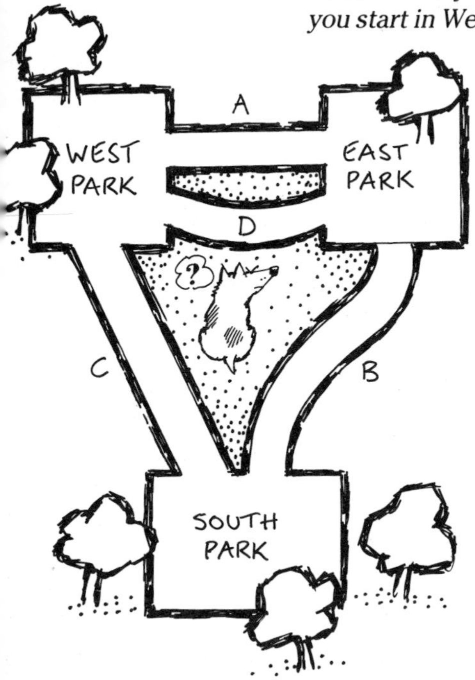

③ Try a slightly more complicated map. Do not use D Road yet. Start from West Park, and go along each road just once. (a) Do you end up where you started from? (b) Does it matter which road you go along first? (c) Does it matter which park you start in; do you always get back to your starting point?

Then imagine that D Road is opened, and you can go along it. Start from West Park, and try going round each of the roads once. Write down your route, and where you end up; for example

West park → A → D → C → B → East Park

(d) Try all the different routes you can find, starting at West Park. Write down each one. Do you always end up in the same place? (e) Now do the same thing, but starting in East Park.

(f) Now try starting from South Park. Can you go round all four roads, once each? Write down each route again. (g) Can you see what makes South Park different from the other two?

④ In the maps above, the parks are called **nodes.** Odd nodes have an odd number of roads leading to them. When there are odd nodes, all complete routes start and end at odd nodes. Can you explain why?

⑤ In the paper-round problem opposite, the nodes are Main Road, the High Street, and the two squares. How many of them are odd?

An extra footpath from a node changes it from odd to even, or vice versa. Can you add footpaths to the paper-round map to make all the nodes even? Now can you solve the original problem?

⑥ Can you work out a general rule about when these network problems can be solved? If someone showed you a new one, how could you say at once whether or not it was possible?

The logic of time

You can measure distance from left to right, or from right to left. But time is different. Time is like an arrow; it goes one way only. We can't measure time backwards, and we can't go back in time.

One logical result of this is that the months and seasons of the year always come in the same order. May always follows April, and April always follows March.

The photographs on these two pages were all taken from exactly the same place, in Hartshead, Yorkshire. For each picture, the ancient camera was balanced on a particular gatepost, and the photographer was up to his knees in stinging nettles.

You may see at once that they weren't taken in the order A, B, C and so on. What can you tell about the order they were taken in?

① How many—and which ones—were taken in winter? Which in summer? Write down how you can recognize the seasons.

② The tree on the right with bright orange berries in picture E is a rowan, or mountain ash. What happens to this tree during the year? What are the berries for? At what time of year do you think they are ripe? What do you notice about the height of the tree?

③ The crop growing in the field behind the rowan tree is barley. When is it harvested?

④ Most of these pictures were taken in the morning. Look at the shadows. Which way do they point in the morning? Make a guess at which way the camera was pointing (north, south, east, west?).

⑤ What can you say about the direction of the prevailing wind?

⑥ In what order do you think the photographs were taken? Explain the logic of your reasoning in at least ten lines.

Murder in the classroom

The morning is cold and dark. Wind howls round the gaps in the doors. Rain lashes at the windows. You go back early at the end of break, walk into your class, and slumped over your desk is a body!

It's Miss Understanding, the maths teacher. There's a knife in her back. The murderer must be one of the three people in the room, Amrit, Baba, and Cathy. Each of them makes just one statement before the rest of the class arrives. This is what they say.

Two of the children are telling the truth, but one is lying.

Can you work out who committed the murder? You don't need to know who had the knife. You don't have to worry about motives. You don't even need to take fingerprints. All you need is logic.

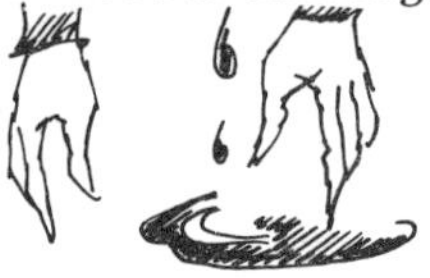

You can apply logic to the murder mystery by making a truth table. Call the suspects A (for Amrit), B (for Baba), and C (for Cathy). Suppose first that Amrit did it, which of the statements was true?

A said he didn't do it; that was false.	*FALSE*
B said A did it; so B was telling the truth.	*TRUE*
C said B did it; so C was lying.	*FALSE*

So you can write F T F *in the first line of the table.*

Which statements are true?	A B C
If A was the murderer	F T F = 1T, 2F
If B was the murderer	T F T = 2T, 1F
If C was the murderer	T F F = 1T, 2F

① This is the complete truth table. So who is the murderer?

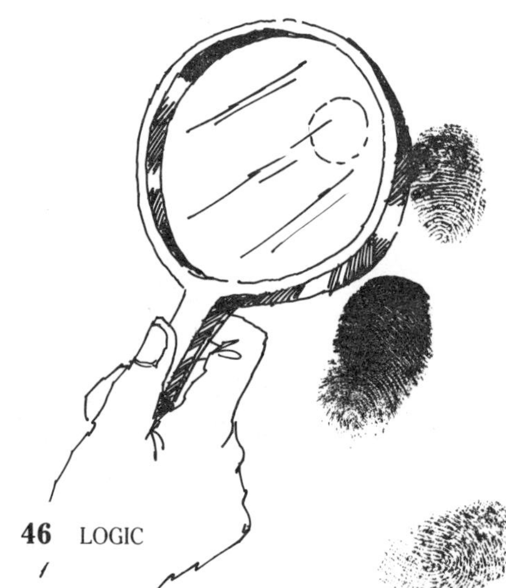

We were told at the beginning that two were telling the truth, and one was lying. The right-hand side of the truth table shows that this is so only if B is the murderer. Therefore we can deduce that Baba committed the crime.

We have been using deductive logic. And with logic like this, there can be no misunderstanding.

Someone has broken a window with a stone. The stone and all the pieces of glass are lying on the floor inside the room.

Where did the stone come from?

There were only three people outside the window who could have thrown the stone. Alice, Beth, and Carl. This is what they said:

But only one of them is telling the truth.

(2) Make a truth table to find out which of the children must have broken the window.

There's another way to find out the answer. Look for two statements that are contradictory.

What Alice says is exactly the opposite of what Carl says. If Alice is telling the truth, then Carl must be lying. If Alice is lying, then Carl must be telling the truth.

So you know that either Alice or Carl is telling the truth.

But you also know that only one person is telling the truth. This one person must be Alice or Carl. You don't know which, but you do know that Beth must be lying.

(3) Beth said that Alice didn't do it. But Beth was lying; so who broke the window?

(4) Look back at the murder mystery opposite. Which two statements are contradictory? Can you use those to work out who did it? Don't forget that in that case two people told the truth. Do you get the same answer as you did with the truth table?

(5) Make up your own murder problem with three suspects who all tell lies. Make sure it is logical, and then try it on your friends.

Oddsocks and her logical frog

Oddsocks hates getting up in the morning when it's cold and dark. She grabs an armful of clothes and runs to put them on in the warm. In her top drawer are three pairs of red socks and three pairs of blue socks. She doesn't care which colour she wears, but she does want a pair — two socks the same colour.

① How many socks does she have to take out (in the dark) to be sure of getting a pair?

The simplest way to tackle this problem is to take a mental step sideways, and approach it from a different direction. Some people call this lateral thinking.

How often can she get it wrong? How many could she take out without getting a pair?

She could take one red, and then one blue. That's two. But the third sock must be red or blue, and so it must match one of the first two. So the answer is that she has to take three socks to be sure of getting a pair.

② Does it matter how many pairs of each colour there are? Write down whether you think it makes any difference, and why.

③ Oddsocks also has a glove drawer, with three pairs of red gloves and three pairs of blue gloves. It's still dark. How many gloves does she need to take out to be sure of getting a left and a right glove of the same colour? Why is this different from the socks?

Oddsocks's pet frog is called Green. He has brothers called Pink and Blue. They like to sit under flowers. By the pond are a green flower, a pink flower, and a blue flower.

None of them sits under the flower that is his own colour, and Pink doesn't like the blue flower.

④ Where does Green sit?

One good way to tackle this is to make a table. Put the frogs across the top, and the flowers down the side.

	PINK	BLUE	GREEN
PINK FLOWER	X		
BLUE FLOWER		X	
GREEN FLOWER			X

You can cross out the spaces on the diagonal, because you know the frogs don't sit under their own colours.

⑤ Copy the table into your book. Where else can you put a cross? Remember that Pink doesn't like the blue flower. So where must Pink sit? When you know, put a tick in the right place.

Now can you finish the bottom line? Remember, only one frog sits under each flower. Then can you see where Blue must sit?

Under which flower will you find Logical Frog Green?

Green and Oddsocks were journeying far, into the Wide Blue Yonder (turn left at Milton Keynes . . .). They stopped at a cave, where the Puzzling Python posed them a question.

⑥ He showed them five cards, face down

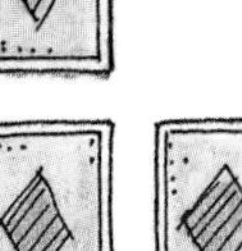

"The Nine lies just beside the Queen.
The King and Queen have a Jack between.
So where's the Ten, Oddsocks and Green?"

Oddsocks worked it out. Can you? Make the five cards from paper. Find a way of putting them down so that they fit the Python's poser. Try the puzzle on your friends.

Make up your own puzzle with the same cards in a different pattern.

The Puzzling Python was so delighted by their logic that he offered to give Green and Oddsocks a treasure chest. There were three old chests on a shelf. The first was labelled GOLD BARS. The second was labelled IRON BARS. The third was labelled MARS BARS.

"You can have any one of the three," said PP. "But I warn you that all three labels are on the wrong chests. You may not look inside, until you get home. All you may do is shake one of the chests, and one only. Then you must choose which one to take."

"Which chest should it be?" said she. "Which one shall we take? Which one shall we shake?"

"No problem!" said Green. "Easy if you use a spot of logic."

Alas here is the bottom of the page; there isn't space for Green's answer. But you can work it out, if you use his sort of reasoning.

⑦ Write a letter to Oddsocks. Explain to her which chest to shake, and how they should choose which one to take home with them.

The mapless sailor

In August 1492
Columbus sailed the ocean blue,
Looking for India, going west,
Because he thought that west was best.

Marco Polo'd made it clear:
"Turn left after Africa."
But Chris ignored him—silly chap.
He didn't have a proper map.

And when he turned up six months later
With parrots and a sweet potato,
The slaves he brought, with skins dark red
Must be 'red Indians' they said.

And so Columbus never knew
Just what he'd found, when all his crew
Picked up their bags of duty-frees,
In what they called the West Indies.

① What land did Columbus discover?

② Write down in two lines what a map is, and in three lines why a map would have helped Columbus.

Columbus had trouble finding volunteers for his trip. Some people said the Earth was flat, and that if he sailed west he would fall over the edge, just as the Sun does at sunset.

③ Write down several reasons why you believe the Earth is round.

Columbus had even more trouble finding sponsors to put up the money for his voyage. One problem was that he could take only enough food for one month at sea.

The Portuguese experts said that even if the world was round, and Columbus found the way to India, it must be a journey of 16 000 km. They reckoned he could sail about 150 km a day. Columbus disagreed: "No, no; it can't be more than 4000 km."

In the end he persuaded the King and Queen of Spain to give him the money for the voyage.

④ (a) How far could Columbus have sailed before his food ran out? (b) How far is it from Spain to India, going west? (c) How far is it from Spain to the nearest land west of Spain? (d) How long do you think the voyage actually took?

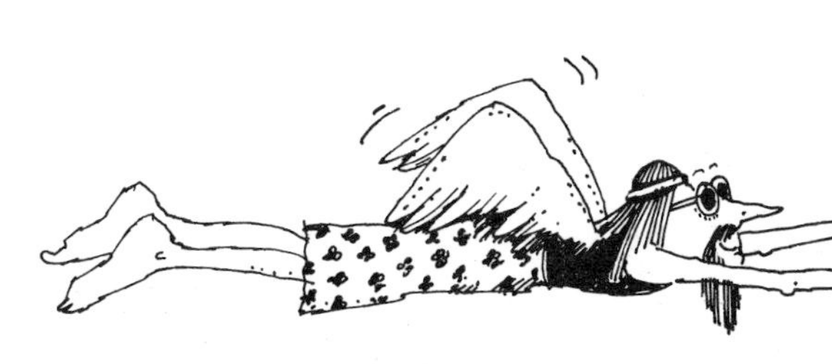

⑤ Eratosthenes had measured the circumference of the Earth some time earlier. Look him up in this book to find out how much earlier. Would it have been useful for Columbus to know about his estimate? Would it have helped Columbus to get sponsorship? Suggest reasons why Columbus might not have known about Eratosthenes and his work.

⑥ Imagine that you were Mathematical Adviser to the Spanish Court. Write a letter to the King or Queen of Spain, giving them advice on whether or not to sponsor Columbus's trip.

⑦ Draw a map to show where you think Columbus actually sailed, and where he landed. Before you start drawing, you might like to find out more by looking him up in the library.

Overground...

These two maps show the same area of London, in different ways.

① Write down five differences between the two maps.

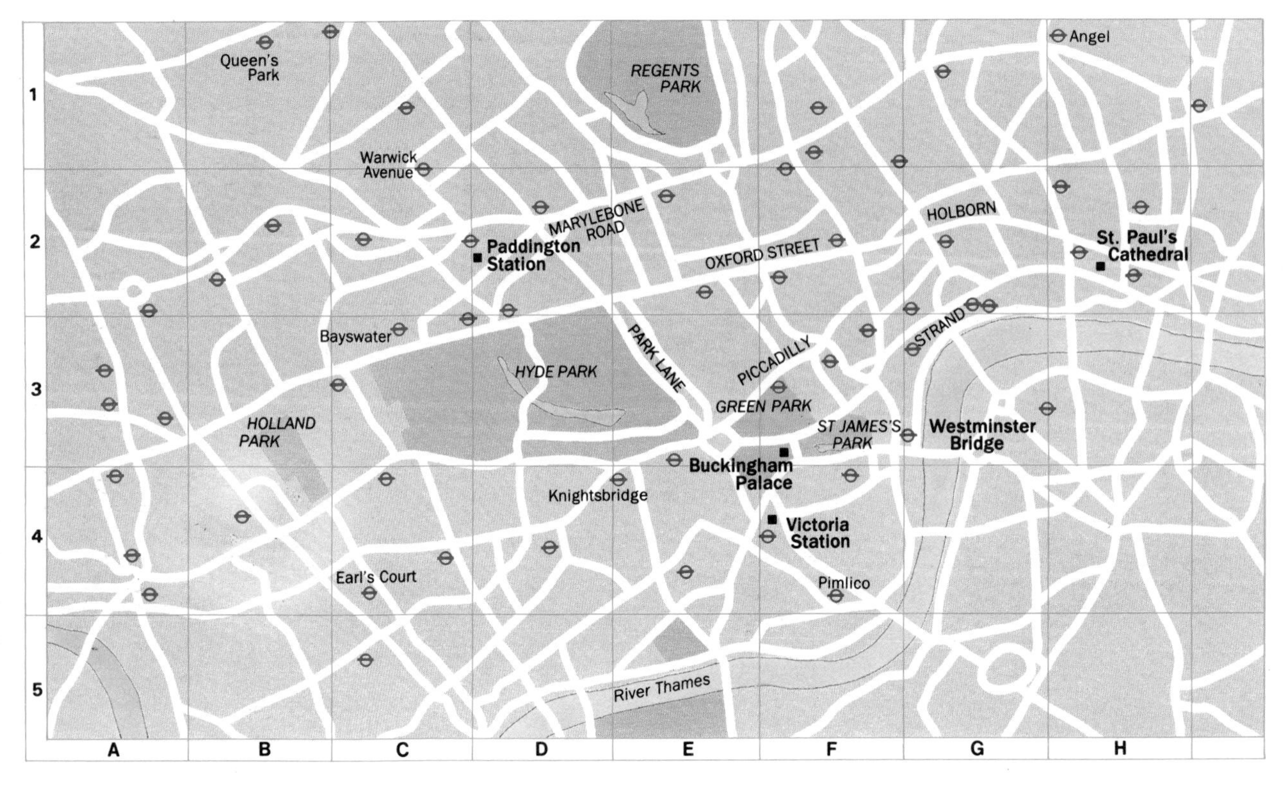

This map on the left is like a photograph taken from high above the ground. It is an accurate scale drawing. Each centimetre on the map represents 50 000 cm on the ground. That's 500 metres.

② How far is it—the shortest distance along the roads—from Paddington Station to (a) Buckingham Palace, (b) St Paul's Cathedral? About how long would each journey take you to walk?

One quick way to find places on the map is to use the blue grid. The squares are numbered A B C . . . across, and 1 2 3 . . . down the side. So Paddington Station is in square D2. Its **coordinates** *are D2.*

③ What are the coordinates of (a) St Paul's, (b) Victoria Station, (c) Westminster Bridge?

④ How far is it across a blue square?

. . . underground

This map represents the London Underground rail network. It doesn't look much like the 'real' map opposite; this map has been stretched and twisted to make the network easier to follow.

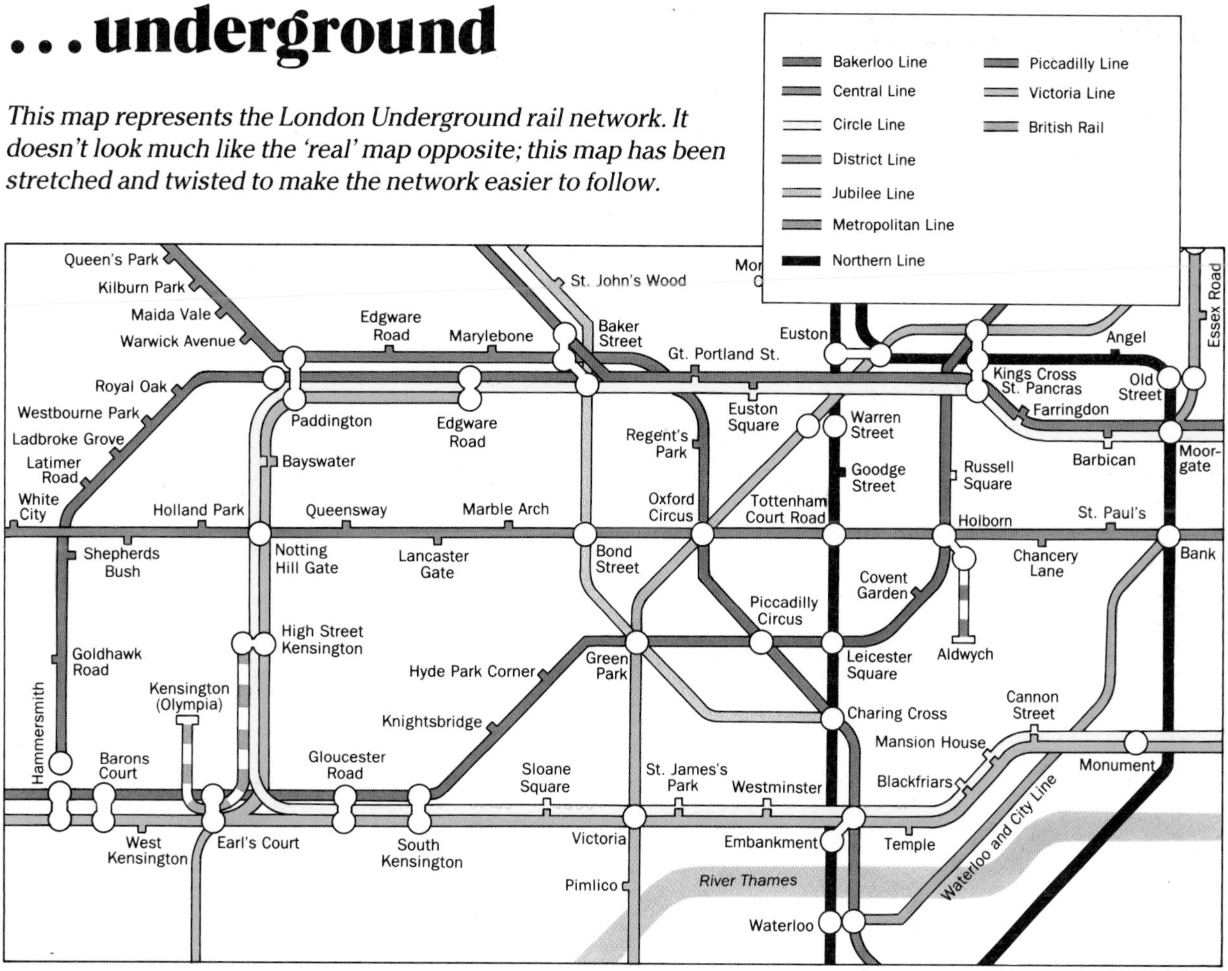

What matters are the connections. When you travel underground from one place to another, you need to know which lines to take and where to change. You don't need to know how far it is, or exactly which directions the trains take, as long as you get to the right place.

Sometimes there is more than one way of getting to a place you want to go. To find the quickest route you can compare journeys roughly like this: count the number of stations, since the train stops at all stations, and add three stations each time you change lines.

⑤ Suppose you wanted to go from Knightsbridge to Victoria (near the middle of the bottom of the map). You could start by going west on the blue (Piccadilly) line to South Kensington, and then east on the green (District) or yellow (Circle). Or you could go east on the blue (Piccadilly) to Green Park, and then south on the light blue (Victoria). Which route would you choose, and why?

⑥ Write a note to a friend, describing more than one underground route from Paddington to St Paul's. How many different lines will each route need? How many stations does each route go through? Say which you think is the best route, and why.

⑦ Find what route on the network has the maximum number of stations you can go through without having to change to another line.

⑧ There are two stations called Shepherd's Bush. What do you reckon is the quickest underground route from one to the other?

GRAPEVINE 1

This map shows the north of Death Valley, the most famous desert in North America. When Frederick V Coville broke camp on 19 January 1872 and entered Death Valley by the old Borax road, he noted the snow-white stretch of salt in the bottom of the valley, and "to the northward, perhaps fifty miles away, mountains, valleys, and salt flats vanished in haze". What he didn't have was a good map.

Every good map has a ***key****, which explains the meaning of the various symbols on the map. It also has a* ***scale****, which tells you what distance on the map represents 1 km, say, on the ground.*

① There are two towns in Death Valley, called Furnace Creek and Stovepipe Wells. Estimate how far apart they are (a) by road, and (b) directly, 'as the crow flies'.

② In how many places can you get food, water, and fuel? Setting off south from Scotty's Castle, how far would you have to go for each of these things?

③ One year, Furnace Creek had no rain at all, but in an average year they get 33 mm. What's the average rainfall where you live?

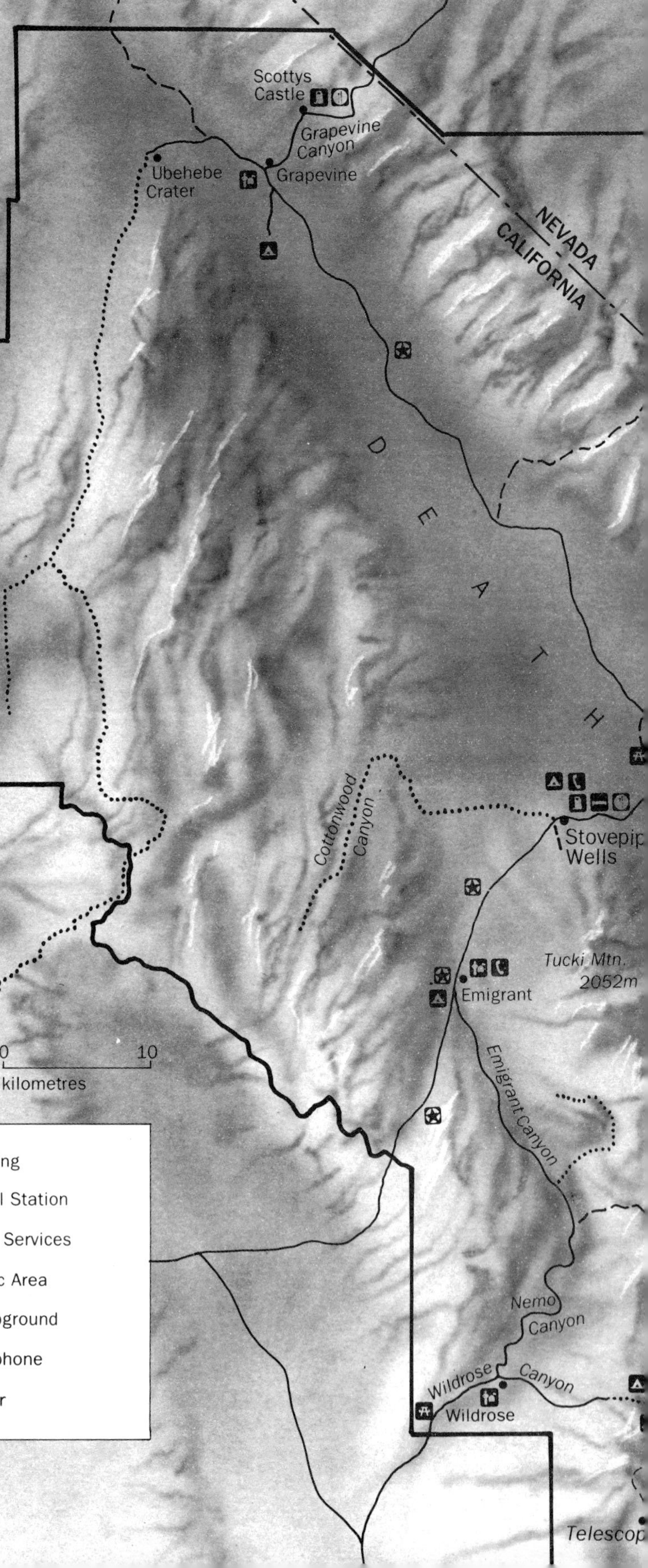

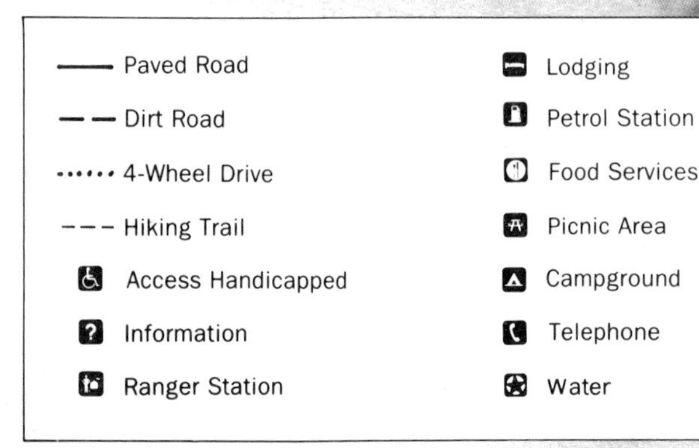

④ Tucki Mountain is over 2000 metres high. What are the highest and the lowest points shown on this map? How far is it from one to the other (a) directly, (b) by trail and road? What is the difference in height between the two?

⑤ Grapevine 1 is the number of the public telephone by the ranger station near Ubehebe Crater. Why do you think it might have been given this number?

⑥ Some of the roads are good, but others are so rough you need 4-wheel-drive vehicles to get along them. List four canyons you could visit in an ordinary car, and four that would need 4-wheel-drive. What do you think a canyon is?

⑦ Make a poster to advertise the attractions of Death Valley as a holiday centre.

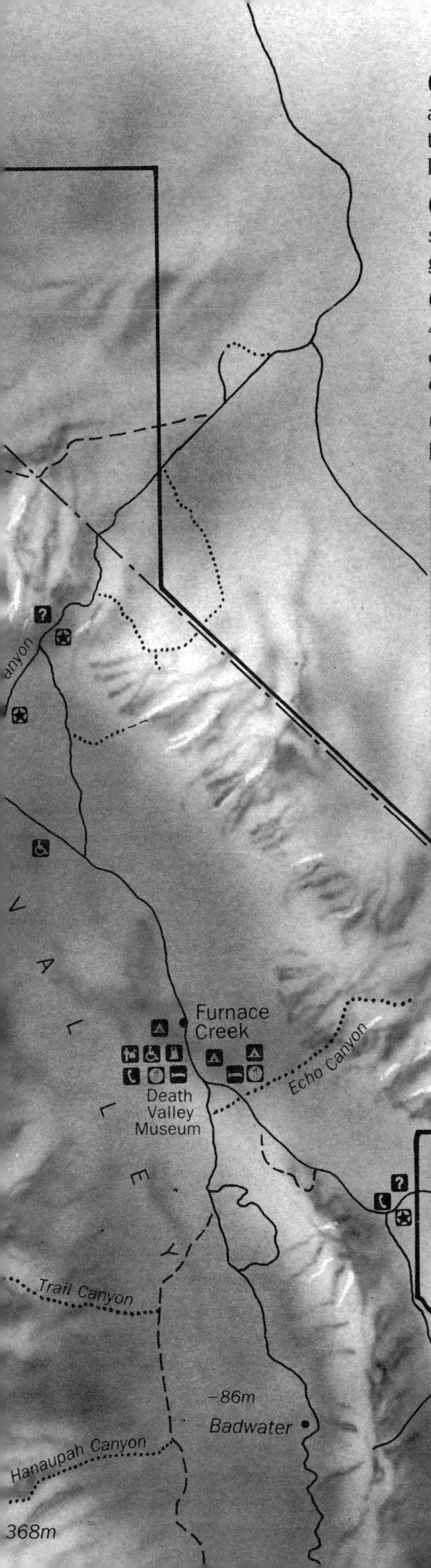

A lengthy mistake

The Trans-Siberian Express is a great lumbering train that trundles across the length of the Soviet Union.

In the middle of the night, it grinds to a halt at Erlian, on the Chinese border, to be taken apart and lifted off the rails. Why? Because of a measuring mistake.

The designers of the world's first trains took their ideas from the trams and carriages that were in use at the time. Many engineers thought the best total width was 5 ft (152 cm), measured from the outside of one wheel to the outside of the other.

① About how wide are cars today, from the outside of one wheel to the outside of the other? How much have things changed in 150 years?

② How many millimetres are there in a metre?

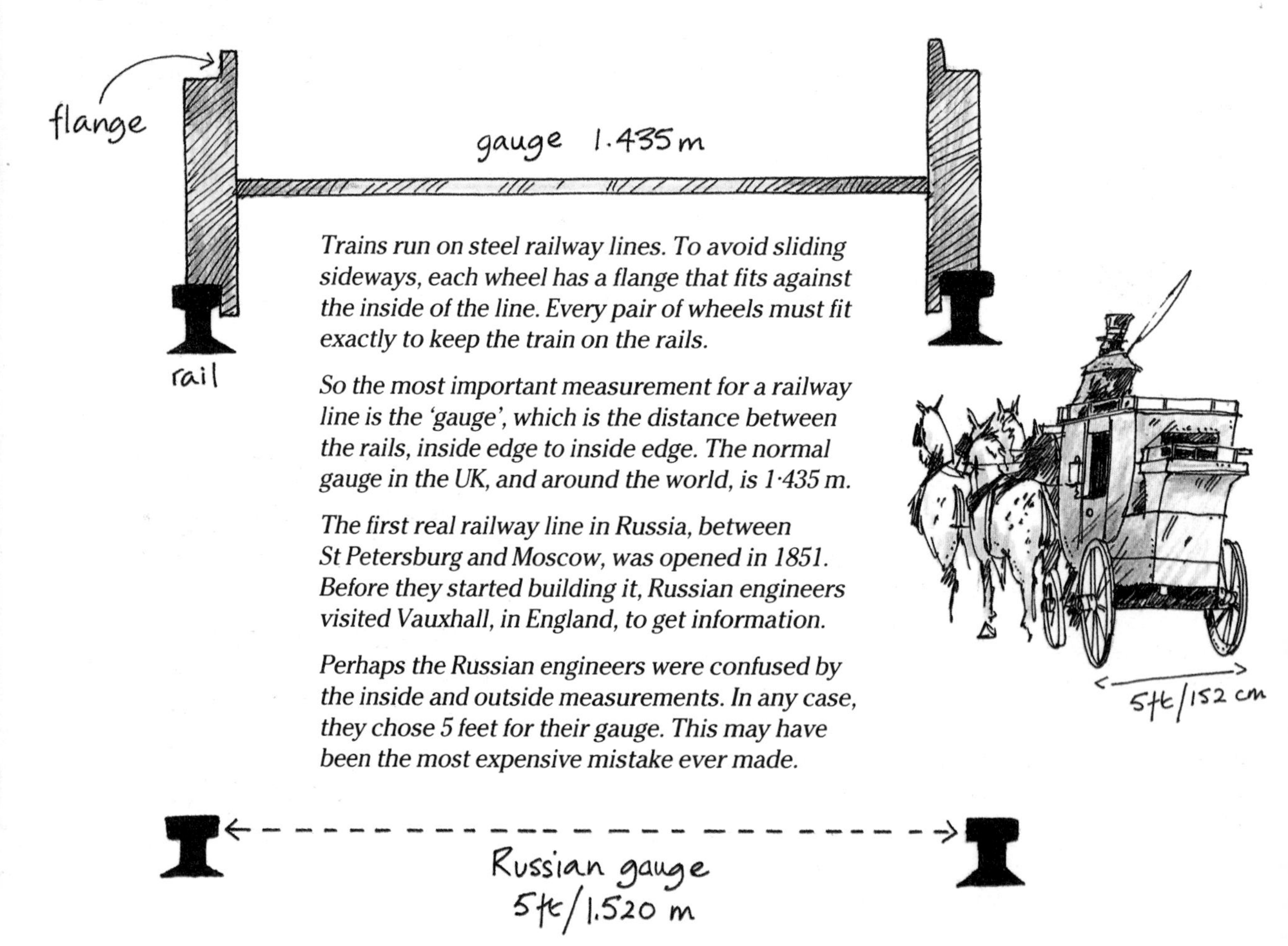

Trains run on steel railway lines. To avoid sliding sideways, each wheel has a flange that fits against the inside of the line. Every pair of wheels must fit exactly to keep the train on the rails.

So the most important measurement for a railway line is the 'gauge', which is the distance between the rails, inside edge to inside edge. The normal gauge in the UK, and around the world, is 1·435 m.

The first real railway line in Russia, between St Petersburg and Moscow, was opened in 1851. Before they started building it, Russian engineers visited Vauxhall, in England, to get information.

Perhaps the Russian engineers were confused by the inside and outside measurements. In any case, they chose 5 feet for their gauge. This may have been the most expensive mistake ever made.

Lengths or distances
1 kilometre (km) = 1000 metres (m)
1 metre (m) = 100 centimetres (cm)
1 centimetre (cm) = 10 millimetres (mm)
1 foot = 12 inches or about 30 cm

Today, the Soviet track has a gauge of 1·520 m. Why does that matter? Because whenever you go from any other country into the Soviet Union you have to change trains—or at least change the bogeys on which the wheels are mounted.

And that is why the Trans-Siberian express changes bogeys at the Chinese customs post at Erlian. Every carriage is trundled into a great shed, and lifted off its wheels by brisk Chinese using huge hydraulic jacks, so that the new bogeys can be slipped in.

This story has a moral. When you are making any sort of measurement, make sure you are measuring the right thing.

③ What is St Petersburg called now?

④ How much narrower is the gauge in the UK than in the USSR? Write the difference in centimetres, and draw a line of that length.

⑤ Look at a map of the Soviet Union. Where do you think the Trans-Siberian Express might start and end its journey?

To estimate the size of something big you can use 'parallax'. Suppose you're looking at a building and you want to know how wide it is. Hold up a thumb towards the building, as far away from your face as you can reach.

Shut your left eye. Line up the left side of your thumb with the left side of the building. Then, keeping your thumb still, open your left eye and shut your right. Your thumb seems to jump across the building. The jump is caused by 'parallax', because your two eyes are in different places.

Suppose the left side of your thumb jumps about twice the width of a door. You know a door is nearly a metre wide. So the parallax jump of your thumb is nearly two metres.

⑥ (a) How can you use this to measure the width of the building?
(b) What happens if you try another building further away?
(c) Can you use parallax to measure height?

Time for a song

One of the first passengers on the train from St Petersburg to Moscow may have been a boy called Peter Tschaikowsky. He was 11 when the line was opened in 1851; his dad was a mining engineer.

PT was red-hot at music, though useless at everything else. His family lived in St Petersburg, but later moved to Moscow.

① In which year was Peter Tschaikowsky born?

In music, the time taken for each note is measured first by the type of note, and second by the tempo of the music. The piece of music above starts with two semiquavers and two crotchets. A crotchet is always four times as long as a semiquaver.

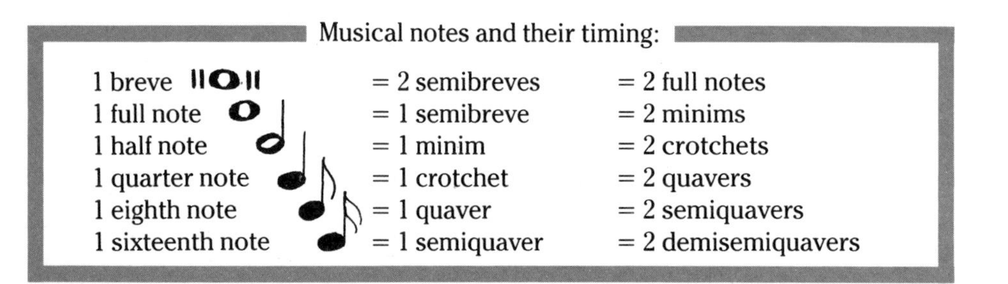

Musical notes and their timing:

1 breve	= 2 semibreves	= 2 full notes
1 full note	= 1 semibreve	= 2 minims
1 half note	= 1 minim	= 2 crotchets
1 quarter note	= 1 crotchet	= 2 quavers
1 eighth note	= 1 quaver	= 2 semiquavers
1 sixteenth note	= 1 semiquaver	= 2 demisemiquavers

② Write down how many semiquavers there are in a full note, and how many demisemiquavers there are in a breve.

At the top of the music for The 1812 Overture *is the word* **Largo**. *This isn't a kind of Russian beer, but the* **tempo** *of the music. Largo means that it's a slow piece of music; precisely, that 60 crotchets should take exactly one minute. So 1 crotchet takes just 1 second.*

③ If the tempo is Largo, how long is a breve? A quaver? A quarter-note? A demisemiquaver?

④ Find out from a music book, or your music teacher, about other musical tempos.

In 1812 the French invader Napoleon had been driven back from Moscow, mainly by the bitter Russian winter.

Sixty years later, the Russians wanted to celebrate the anniversary of Napoleon's defeat; so they invited Peter Tschaikowsky to write a piece of music. What he produced was The 1812 Overture.

⑤ When was *The 1812 Overture* written?

PT had no way of making records, tapes, or compact discs. But he wrote his music down, and we can still enjoy it today. It is noisy and dramatic; full of marches and bells and cannons firing. Listen to it some time.

⑥ Copy and complete this table of units of time:

1 century = 100 ____s	1 day = __ hours
1 decade = 10 ____s	1 hour = 60 ______s
1 ____ = 12 months = about 365 days	1 ______ = 60 seconds

⑦ (a) How many seconds are there in a year?
(b) How long would a million crotchets take, if the tempo was Largo?
(c) How many times do you think you have breathed in your lifetime?

⑧ Suppose you were chief cook on the train from St Petersburg to Moscow. A royal couple ask for boiled eggs for breakfast. The Princess would like her egg boiled for 8 minutes; the Prince wants his done for 4.

You have no clock or watch, but you have two egg-timers. One of them measures 3 minutes; the other measures 5. How can you use them to measure the right times for the eggs?

What other times could you measure with these egg-timers?

High church

The largest wooden church in the world is at Kerimaki, in eastern Finland. Standing on a hill, it can be seen for miles and miles. There is room inside for 3400 people—many more than are ever likely to go there, since Kerimaki is only a village.

The story goes that when they were planning their new church in 1842 they sent to England for information, because they thought English architects were good. The plans arrived, and all the measurements were in feet. Unfortunately the Finns thought they were in metres.

① How many feet are there in 1 metre?

② How many times taller is the church than it was meant to be?

③ Suppose you had to clean the floor in the church. How many times bigger is the area of the floor than it should have been?

④ Finland has cold winters. Kerimaki is just outside the Arctic Circle. Heating all the air inside the church costs much more than it should have done. By how much has the volume of air inside been multiplied because the church was built in metres instead of feet?

When you are just counting, all you need is numbers. To find out how many churches there are in Finland, or how many people go into the one at Kerimaki, you need only numbers.

⑤ How many legs would you expect to find on a biped, a tripod, a quadruped, a centipede, and a millipede?

To count legs you need only numbers. But to measure their length, or the time, or a mass, you need more than just numbers. You also need ***units:*** *metres, or minutes, or milligrams, for example.*

What went wrong with the Finnish church was the units. Feet and metres are units, and if you want to build anything, or measure anything, you have to understand how units work.

Suppose you want a new door; it's no use going into a door shop and saying "I want a door about three wide". Three what? Metres? Grams? Minutes?

⑥ Write down a sensible measurement for the width of a door. Don't forget the units.

Suppose you're cooking a meal. You're in trouble if the recipe says "Take 3 of rice". Three grains? Three sacks? Three tonnes? Three decibels? You must know the units.

⑦ Write a part of a recipe that includes 3 ___ of rice, 3 ___ of butter, 3 ___ of milk, and 3 ___ of curry powder. Use different, sensible units for each.

(8) Match up the things you might want to measure with the units you might use:

Musical note	Metres (m)
Mass	Millimetres (mm)
Thickness	Hours (h)
Time	Metres per second (m/s)
Distance	Kilograms (kg)
Speed	Kilometres (km)
Height	Quavers

(9) Find out and write down what these units are used for:

cubic centimetres (cc), decibels (dB), degrees Celsius (°C), goals, holes, light years, litres (l), millibars (mb), m.p.h., and volts (V).

(10) Mr Segg has scrambled his shopping list, again. What it says is 500 litres of string, 2 dozen tomatoes, 1 metre of milk, and 50 grams of eggs. Knowing him, he's probably got everything mixed—the numbers and the units. What do you think he really wants? Can you explain why? (Hint: Think about units, then about sensible amounts.)

(11) What does 3.25 mean? It might be room 25 on the third floor of a hotel. Write down at least two other things it might mean if there were no units to tell you.

Now write 3.25 with suitable units to make it: (a) walking speed, (b) the length of a pop song, (c) the depth of a swimming pool, (d) the height of an ant, (e) the distance to the nearest railway station, or (f) the speed of a bullet.

(12) Go to a shop or supermarket and see how many different units you can find being used on things for sale. Write them all down, with their abbreviations; then sort them into groups. What relationships can you find between different units? What other units can you think of that you can't find in the shops?

The swinging estimator

Galileo Galilei was a brilliant Italian scientist and mathematician. He never let himself get bored. Whenever his life began to look boring, he used his imagination.

One day in 1583 there was a mind-bendingly boring sermon in Pisa cathedral. Everyone else went to sleep, but Galileo watched the new lamp, and the way it was swinging. He wondered . . .

Does every swing take the same length of time?

Does it swing faster when it isn't swinging so far?

Would it swing more quickly on a shorter chain?

Does the weight make any difference?

① What do you think? Write down what you think are the right answers to Galileo's questions.

To answer his questions, Galileo needed to measure some times. Digital watches hadn't been invented; so he used his pulse.

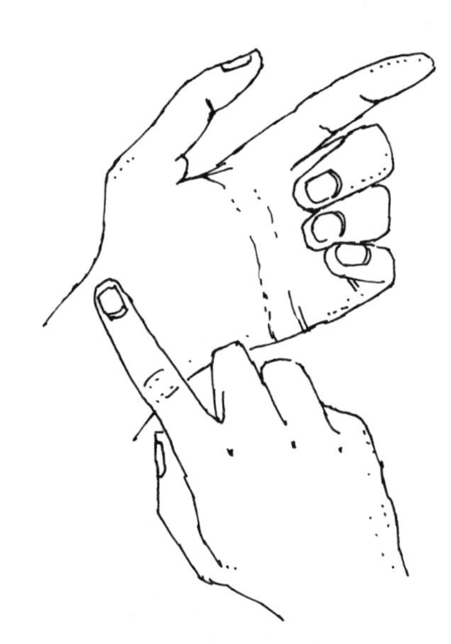

Find your own pulse by feeling with your first finger either just above the bone on the inside, thumb-side, of your wrist; or at the top of your neck, half way from your chin to your ear.

Do this next bit only if you have been sitting down for five minutes. If you have been running around, your pulse won't be steady. Galileo had to sit still for hours in the cathedral; so his pulse was steady.

② Find out how steady your heart beat is. Use a digital watch or stopwatch. Write down the number of beats in (a) 30 seconds (b) 1 minute (c) 1 minute 30 seconds (d) 2 minutes.

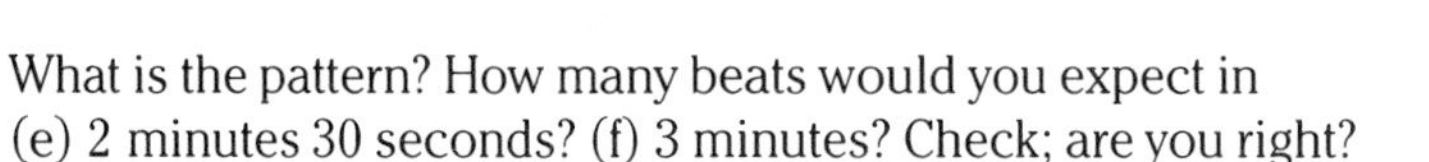

What is the pattern? How many beats would you expect in (e) 2 minutes 30 seconds? (f) 3 minutes? Check; are you right?

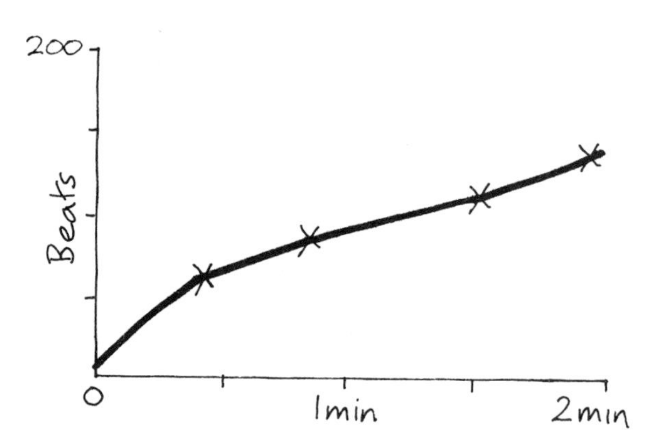

③ Draw a graph of the number of your heartbeats (up the side of the paper) against time (along the bottom). What shape should the graph be if your pulse is regular? Is your pulse regular enough for you to be able to time things with it?

④ Suggest a reason why music written in the tempo **moderato** (see page 58) has 72 half-notes in 1 minute.

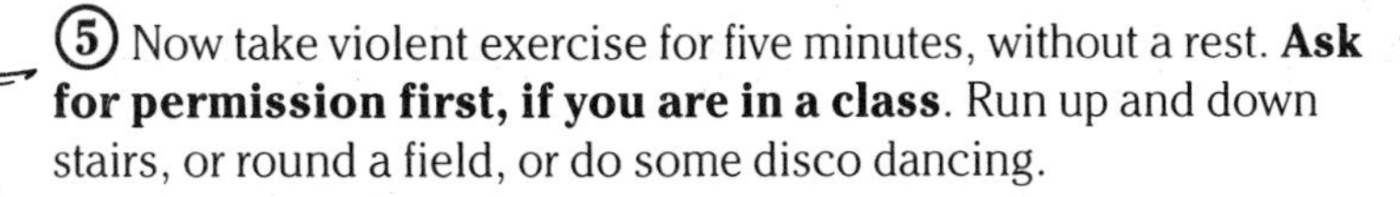

(5) Now take violent exercise for five minutes, without a rest. **Ask for permission first, if you are in a class**. Run up and down stairs, or round a field, or do some disco dancing.

Then time your pulse again, and see what has happened to it. How does it change during the next five minutes, while you sit still?

I have estimated the time of the swing for everything in the church that swings.

. . . The sermon was still going on, but Galileo didn't go to sleep. Instead, he thought about what he had noticed.

I have checked the angle of the swing, the length of the string and the weight on the end.

As far as I can tell, only one of these things has any effect on the time of the swing.

(6) Make a pendulum by hanging a weight on a piece of string. Measure how long it takes for ten small swings. Then try ten big swings. Try varying the length of the string, and the weight.

Write down the time for ten swings in each case.

What do you notice? What affects the time?

(7) Write a note to Professor Galileo Galilei at the University of Pisa. In ten lines explain what you have found out about pendulums, and how long they take to swing.

Headlines

Look at these newspaper stories. They all have numbers in them. Some are exact numbers. Others are roughly right, but not exactly; they are approximate.

1. Which of these numbers are exact? Write down a list.
2. Write down a list of numbers that are approximations.
3. Are any just guesses?
4. Why don't all news stories give you exact numbers? Would it be better if they did?
5. Can you explain when approximations might be better than exact numbers?

Yellow fever

How many buttercups are there in this field? You could count them all, but . . .

① How would you know you hadn't missed one?

② How could you **guess** how many there are?

③ How many do you think there would be under this box?

④ How would you guess the number of blades of grass in the field?

⑤ If you wanted to kill the buttercups and each needs 1 ml of weedkiller, what volume would you need altogether? How much would it cost, if the weedkiller costs £2 a litre?

⑥ Go out and estimate how many buttercups or daisies there are on a piece of grass near you.

Guessing games

Every week people eat a lot of food and use up other things in the house. Soap. Toilet paper. Light bulbs. Every week someone has to go and buy more stuff. How much of each thing should you buy? How many light bulbs? How much soap? How much food?

You can't tell, exactly. You have to guess, or estimate.

① If you had to do the week's shopping how would you estimate how much stuff to get?

Cooks need to estimate when they prepare meals. They guess how long the process is going to take, so they know when to start. Often they estimate amounts such as a cup of milk or a teaspoonful of spice. They might turn the oven a bit higher than the recipe suggests, because the food has been in the fridge.

② A cook cuts 25 g of butter, following the line on the wrapper. Why is this a guess? How could you measure 25 g of butter accurately?

③ Explain why cooks don't need to measure everything accurately, and why estimates are often near enough.

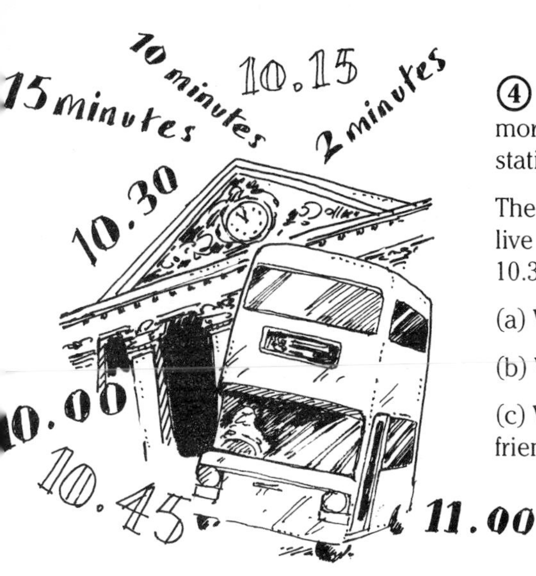

④ You have arranged to meet your friend in town on Saturday morning, at the town hall at 11.00. You can go by bus to the bus station, and the bus ride takes about fifteen minutes.

The town hall is about ten minutes' walk from the bus station. You live two minutes from the bus stop. The buses leave at 10.00, 10.15, 10.30, and 10.45.

(a) Which bus should you try to catch?

(b) What time should you leave home?

(c) What guesses do you have to make if you are going to meet your friend on time?

In a shop you decide to buy a book for £2·95, a record for 99p, a newspaper for 24p, and a magazine for 70p. You need to know roughly what the total is.

However you work this out, the first thing to do is to ***estimate****— that is, make a sensible guess.*

⑤ £2.95 is a bit less than £3, and 99p is just under £1. The newspaper and the magazine together are going to be about £1. So the total must be a bit less than £3 + £1 + £1, or a bit less than how much?

Guessing is very important when you're shopping.

⑥ Suppose you pay for your book, record, magazine, and newspaper with a £10 note, and you get 12p change. Has there been a mistake? About how much change do you expect?

Some people think that if you have a calculator you don't need to bother with this sort of guessing. Wrong! Guessing, or estimating, is just as important if you use a calculator. Maybe even more so!

Calculators cannot stop you making mistakes. Use a calculator wrongly, or press the buttons carelessly, and you'll make a mistake with no trouble at all. The problem comes if you always believe what your calculator tells you, without checking it.

If you enter 2·95 + 99 + 24 + 70 into your calculator, the answer will be 195·95. That's nowhere near £5; so your estimate tells you that you have made a mistake—as long as you make an estimate!

⑦ How did the error happen?

Making mistakes is easy, whether or not you use a calculator. The difficult thing is to spot them—unless you have already guessed the answer. That's why estimates matter.

The numbers game

This is a game for two players or more.

You will need a die, a piece of paper each, and a pencil or pen.

The object of the game is to score the highest two-digit number.

Mark your paper with a double box, like this: ☐☐
This will become your two-digit number.

Throw the die. Whoever throws the higher number plays first. Play then goes clockwise round the circle of players.

The first player throws, and writes the score in one of the boxes. This number must be written down before any other throw is made.

The second player then throws, and writes down the number scored.

Each other player (if there are more than two) has one throw before the first player throws again.

Example game.

	Tens	Units
Your brother gets first throw, and throws a 3. He writes it in the units box:		3
You throw a 4 and write it in the tens box:	4	
He throws a 5 and puts it in the tens box:	5	3
You throw a 6 and have to put it in the units box:	4	6

You lose, because 53 is a bigger number than 46.

What makes this game tricky is that you have to chose where to write the first digit before you see the second.

① Write down the highest digit you could score in one throw, and the lowest.

② Write down the greatest and smallest numbers you could score in two throws.

Try playing the game a few times—either against friends or on your own. You may find you are not always sure which square to write the first digit in.

③ Write down which digits are easy to place, and which are tricky. Explain why.

Slippery Susie cheats at this game. If she realizes she has picked the wrong box she may try to turn the paper round; so the first box becomes the second. But when she does that the digits turn over.

④ Which digits from 1 to 9 still look like proper digits when the paper is turned round? Write down as many two-digit numbers as you can that still look like proper numbers when you turn them over. Put them in two groups—the ones that turn into different numbers, and those that stay the same. Can Susie's method of cheating ever work?

You may even be able to find some words that do this . . .

⑤ Try playing the game using three digits instead of two. Does this make it easier or harder? What if you used four digits, or five?

⑥ Your friend has been chosen to represent her class at the numbers game, playing with three digits. Write her a note with one or two tips on how to increase her chance of winning.

Nothing to it!

Zero is a strange number. What does it mean? Nothing at all! Zilch. Nil. You don't use zero to count with. Try counting your fingers; which one is number zero?

Have you ever seen a number 0 bus? Or a house numbered 0? Or anything else labelled zero?

You might think zero would be useless. But without zero, writing large numbers is difficult. The Romans did not realize the value of a sign for nothing; so their numbers became much clumsier than the Arabic numerals we use today.

Roman numerals	I	V	X	L	C	D	M
Arabic numerals	1	5	10	50	100	500	1000

The Roman system looks neater, but numbers in between these had to be built up by adding. There was no symbol for two; so they had to add I and I. So II = I + I = 2, III = 3, VI = 6, VIII = 8, XII = 12, MMI = 2001, and MMMDCCCLXXXVIII = 3888.

① In the women's singles final at Wimbledon, the game begins fifteen–love, thirty–love, thirty–fifteen. Write down these scores in numbers. Can you do it in Roman numerals? Why not? What games could you score with Roman numerals?

When you are writing Roman numerals for numbers ending in 4 or 9, it's quicker to subtract. For 4 you can write IV instead of IIII, because IV means '1 less than 5'.

So IX = 9, XIV = 14, XIX = 19, IC = 99, and MIM = 1999.

I bought an old book with MDCCCLXXIV on the title page. When was it published? Take apart the Roman number:

M | **D** | **CCC** | **L** | **XX** | **IV**

$$1000 + 500 + (100 + 100 + 100) + 50 + (10 + 10) + (5 - 1) = 1874$$

② Write down in Roman numerals (a) your age, (b) the number of people in your class, (c) the date, month, and year.

Roman numbers are hopeless for calculations. Just try adding IV to IX. Multiplying and dividing are even worse.

What saved the world was the brilliant Indian invention of the zero.

The zero is important because it lets us write all numbers in columns of units, tens, hundreds, and so on, even if there's nothing in some of the columns.

Think about the numbers 21 and 201. You know 201 is a greater number because there is a 2 in the hundreds column. But you know that only because of the zero. Without the zero, 201 is the same as 21.

In other words the value of a digit depends not only on its face value, but also on its place in the number.

③ Write down the number 21. Then write down all the numbers you can get by adding a zero at the front, in the middle, and at the end of the number. Which is the greatest?

④ Write down the numbers MC, CX, XI, MI, MX, and CI in Arabic numerals. Put the Arabic numbers in increasing order of size, and explain how you know how to order them.

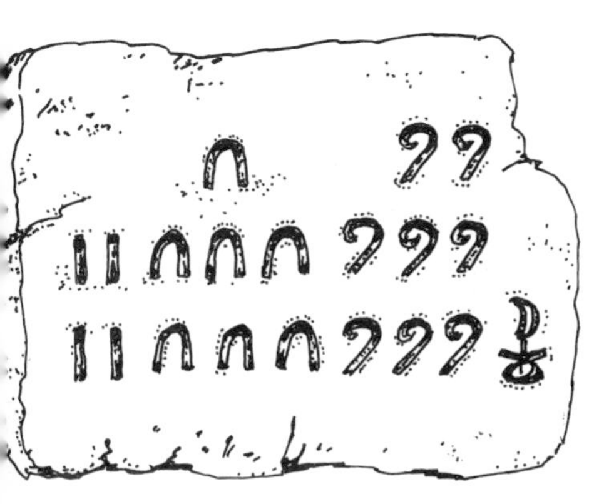

People have always counted on their fingers. That probably explains why we call single numbers digits, since digit means finger. And it probably explains why we count things in tens.

⑤ What sorts of things do we count in tens? Money? Lengths? What about time, or days and weeks and months? Make a list of things we count in tens and things we count in other numbers. For the second list, write down what numbers we use.

⑥ This is how ancient Egyptians would have written the number 1874. Do you think they had a symbol for zero? Did they use a place-value system like ours? Write the date in Egyptian symbols.

Below zero

This man seems to be gazing out to sea. In fact, he is 86 metres below sea level. The water is not sea, but a salt pool called Badwater in the middle of Death Valley.

In the south-west corner of the United States the Pacific Ocean washes the shore of California, keeping the coastline green and beautiful. But there isn't much rain, and just a few kilometres inland the desert begins.

Near the border of the next state, Nevada, lies the most famous bit of this desert, Death Valley *(see page 54)*.

In the middle of the valley is a spring. The water from this spring is so salty that the pool never dries up, even though Badwater is one of the hottest places on Earth.

(1) Suggest reasons why Death Valley and Badwater might have been given those names.

Badwater is the lowest bit of land in the United States. It's actually below the level of the Pacific Ocean. There's a sign that says SEA LEVEL *86 metres up a cliff.*

Pacific Ocean

Heights of things are usually given in metres ***above*** *sea level. The top of the Empire State Building is about 300 m above sea level. The top of Mt Everest is about 8800 m above sea level. The highest mountain in Death Valley is Telescope Peak, 3368 m above sea level.*

(2) How far would you have to go down from Telescope Peak to Badwater?

(3) People write the height of Telescope Peak as a positive number: 3368 m means +3368 m. How could you write down the height of Badwater above sea level?

Divers and submarine captains need to know their depth—in other words their negative height above sea level.

(4) Who else do you think might need to know their negative or positive altitudes? Write two lines about each of three people, and about why they need to know.

(5) Imagine that a horrendous saltbeast slithers out of the Badwater pool at midnight, and starts up the cliff. Every hour it crawls up 20 m, and then, while it rests for the next hour, it slips back 10 m.

At what time will the saltbeast reach sea level? (Hint: Write down its height at 1 a.m., 2 a.m. etc.)

(6) Suppose you were a mining engineer, and had dug a tunnel all the way through California, from the pool at Badwater to the Pacific Ocean. You have put a door across each end of the tunnel.

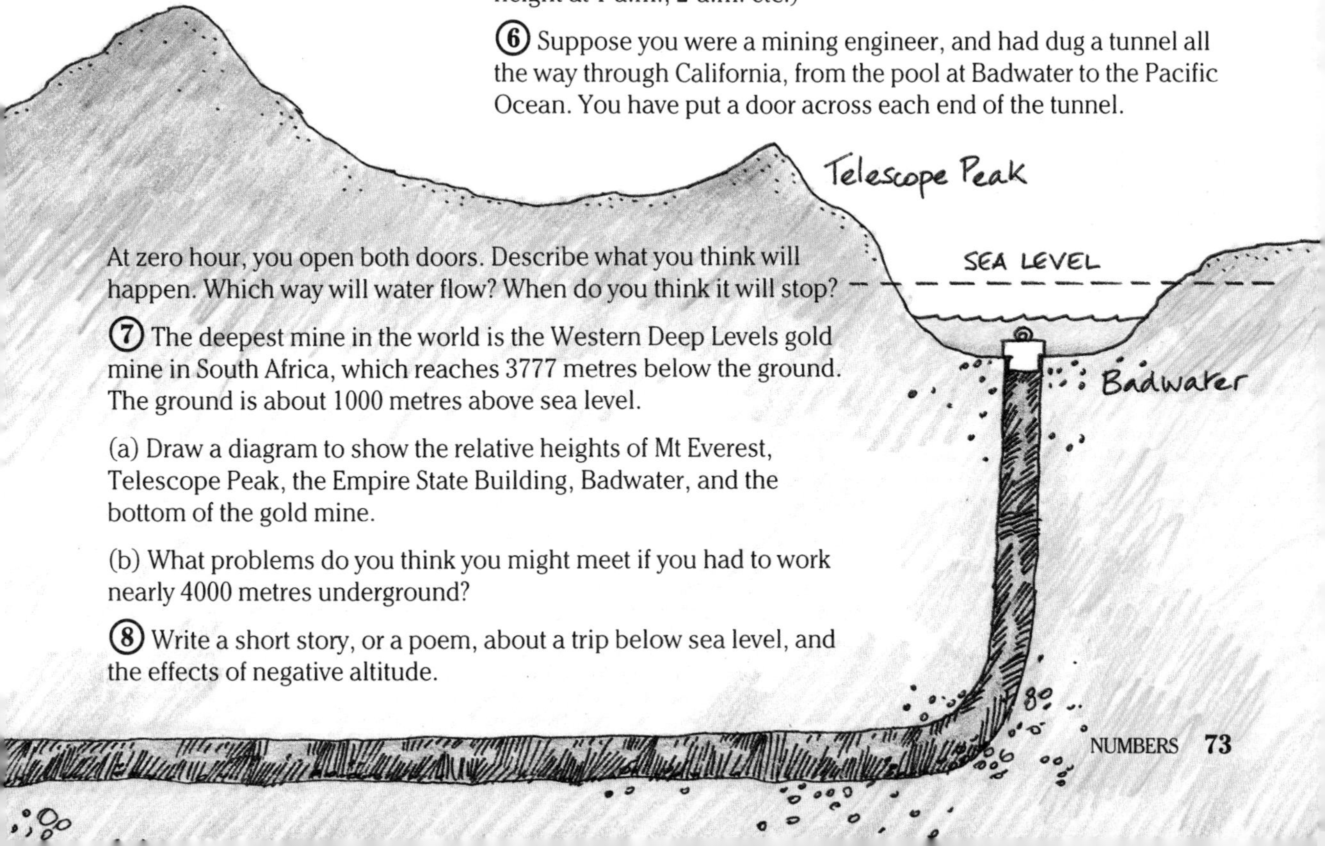

Telescope Peak

SEA LEVEL

Badwater

At zero hour, you open both doors. Describe what you think will happen. Which way will water flow? When do you think it will stop?

(7) The deepest mine in the world is the Western Deep Levels gold mine in South Africa, which reaches 3777 metres below the ground. The ground is about 1000 metres above sea level.

(a) Draw a diagram to show the relative heights of Mt Everest, Telescope Peak, the Empire State Building, Badwater, and the bottom of the gold mine.

(b) What problems do you think you might meet if you had to work nearly 4000 metres underground?

(8) Write a short story, or a poem, about a trip below sea level, and the effects of negative altitude.

The slicing of the cake

Granny's hundredth birthday. Her centenary. The whole family came to celebrate, and she had a huge family; 99 relatives.

The cake was the biggest they had ever seen in the town of Partspur Cent in South Wales. They cut it into ten fat wedges, one for each branch of the family. Each branch of the family had ten members.

They cut each wedge into ten slices; so everyone at the party could have one slice.

① How many slices did they have to cut the cake into altogether?

Ten slices of the cake was ten out of one hundred, or ten per hundred, or ten per cent, i.e. 10%

Per cent is from the Latin word centum *meaning hundred. You can write 10% or ten per cent, because % means per cent.*

② What does 10% mean, both in terms of a part of a whole thing, and in terms of slices of a cake?

HAPPY BIRTHDAY GRAN

At some point in the afternoon, all the children in the family collected in one place. There were 36 altogether.

③ What percentage of the cake did the children get? What percentage did the adults get? What is the link between these percentages?

④ Twenty of the children were girls. What percentage of the family was boys?

Value Added Tax—usually called VAT—is an extra amount added to the costs of all goods and services.

⑤ Value Added Tax is charged at a rate of 15 per cent. If the cooks for granny's party charged £100 + VAT, what was their total bill? The cake alone cost £20 + VAT. How much was the VAT on the cake?

⑥ Is a circle the best shape for a huge cake? And is this the best way to cut it? Suppose you had to make a cake for 100 people. What shape would you make, how would you cut it, and why?

⑦ Write down what you think a centenary is, and the reason for the name. How many cents are there in a Canadian dollar? Think of three other words that come from the Latin word **centum**. Write down what each one means.

Testing for truth

I bought these things in my supermarket. Both claim to have 15 per cent more contents than usual, and both have a red part, suggesting that it corresponds to the 15 per cent.

① Draw a table in your book; one column for toothpaste and one for lemonade. Use a ruler to measure the usual length or depth of the contents and the *extra* amount. Write these values in your table.

② Divide the extra amount by the original amount. You should get a decimal number beginning 0·1. This is the ratio of the extra to the original. Write down the ratio for each product.

The ratio is like a percentage; it shows what decimal part of the original packet is extra. To turn the ratio into a percentage multiply it by 100. A simple way to do this is to move the numbers two places to the left:

Ratio	0·10	0·12	0·155	0·25	0·8	1·25
Percentage	10	12	15·5	25	80	125

③ Write down the 'extra' percentage in each of the two products, according to your table.

FRACTION	DECIMAL	PER CENT
1/2	0.5	50
1/3		
1/4		
1/5		
1/6		
1/7		
1/8		
1/9		
1/10		

Suppose the extra amount was one quarter of the original. The ratio would be $\frac{1}{4}$.

④ Write down $\frac{1}{4}$ as a decimal, and as a percentage.

⑤ Make three columns. Write down the decimals and percentages that correspond to each of these fractions: $\frac{1}{2}\,\frac{1}{3}\,\frac{1}{4}\,\frac{1}{5}\,\frac{1}{6}\,\frac{1}{7}\,\frac{1}{8}\,\frac{1}{9}\,\frac{1}{10}$. Which fraction is closest to the extra amount of each of the products above?

⑥ Are the manufacturers fair in their claims to have added 15 per cent extra? Have you enough information to be sure? Can you think of a better way of finding out the truth?

⑦ Carry out your own investigation to see whether a 'percentage extra' claim is true. If not, write to the manufacturer and explain why not. But make sure you have done your calculations right first!

Vote for Brown!

The Green party believe in ecology. The new Brown party believe in deserts, rust, and chocolate pudding. What chance have they of winning the next election?

Newspapers often use pie charts to show the results of election surveys. Suppose a newspaper asks 100 people "Who are you going to vote for?"

Shown as a pie chart their results might look like this.

The whole pie represents the 100 people asked.

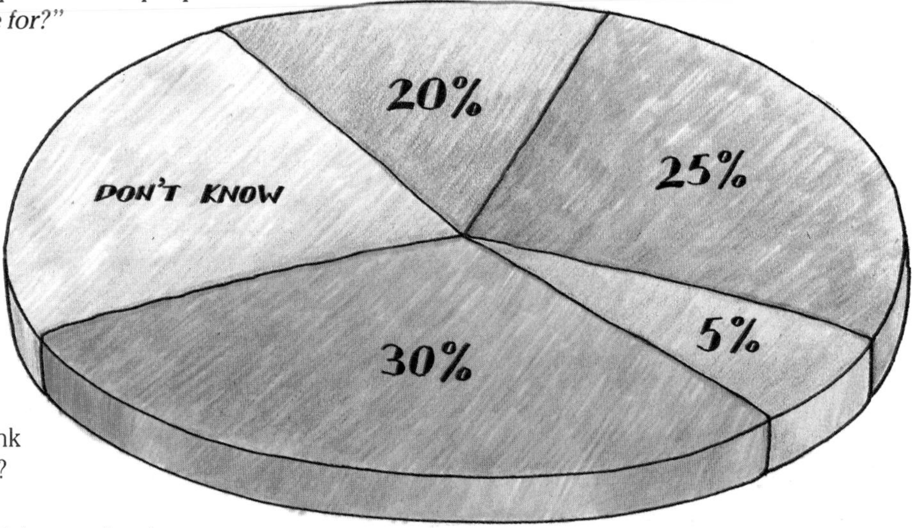

DON'T KNOW
20%
25%
5%
30%

① Which party do you think is going to win the election?

② How many people didn't know who they were going to vote for. What percentage is this of the whole?

③ What fraction of the people say they will vote Brown?

④ Suppose all the people who say they don't know decided to vote for the Green party. What difference do you think that would make to the result, and why?

Pie charts are useful for comparing parts of a whole. They give you a good visual impression of the different percentages that make up the total.

⑤ If you asked everyone who could vote, instead of just 100 people, do you think it would make any difference to the results? Is it fair to ask only 100 people? Why don't they ask everyone?

⑥ Fifty per cent of a cake or pie is exactly $\frac{1}{2}$. Ten per cent is $\frac{1}{10}$. Find out what other percentages make simple fractions. Draw up a table to show the relationships between percentages and fractions.

WE'LL GIVE YOU DESERTS, RUST & CHOCOLATE PUD
VOTE BROWN
BROWN

Three per cent off

Everyone likes getting a bargain. So when a shop announces a sale, people are interested. But watch out! Some sales don't give you such good bargains as you might think.

Most sales seem to offer a percentage discount—a few per cent off the original price. The first step towards making sure the bargains are genuine is to understand how percentages work.

① What is one per cent of £1?

② A giant packet of Doggibrek usually costs £1. Today everything is being sold at three per cent off. So what does a giant packet of Doggibrek cost today?

The usual price is £1, or 100p. That is 10 × 10p.

Three per cent off means the new price is 3 per cent less.

③ The large packet of Doggibrek normally costs 66p. With three per cent off, what is the new price?

④ Write down what a giant packet of Doggibrek would cost if the supermarket had taken off not 3% but (a) 5% (b) 10% (c) 22%.

⑤ Cattosludge usually costs 50p, Rice'n'easy £1, and Liquid-K-2 £2. Write down what the new price of each would be if the supermarket took ten per cent off.

⑥ If you bought two packets of Doggibrek at a special price of 50p each, and sold them to your worst friend for 75p each, what profit would you make? What percentage profit?

⑦ A hard-headed lady called Sade
Charged 12p a can for limeade.
Her profit, that meant,
Was twenty per cent.
So how much a can had she paid?

⑧ At the winter olympics for fleas,
The world record high jump on skis
From a metre was raised
By an eighth! I'm amazed!
How many per cent is that, please?

⑨ A student found somewhere to stay
For just one pound fifty a day.
But they put up his rent
By two hundred per cent;
So what will he now have to pay?

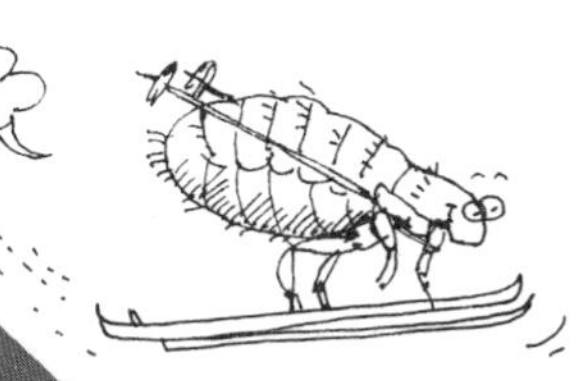

⑩ You buy a shop-soiled game of Swindle for 20% off its normal price of £1. How much do you pay? You clean it up and sell it to a wally for £1. The wally moans that you have made 25% profit. Is that right? Can you explain why? Where has the extra 5% come from?

Draw up a table about the Swindle game to show how much profit you would make if various percentages were knocked off the usual price. Can you work out a rule to explain what happens?

3% less white paper

Jackpot!

In July 1981 Jeff Randolph pulled the handle of a fruit machine in Caesar's Palace, Lake Tahoe, and won a million dollars.

Jeff claims he knew he was going to win. On the way up in the plane he shaved for the second time in the day. He had never done that before. He did it because he "wanted to look extra nice for all the pictures that were going to be taken, after I won the money".

At Caesar's Palace he had a good dinner, wandered into the casino, and began calmly feeding dollars into the one-armed bandit. Twenty minutes later he hit the jackpot. $992 018 said the machine, and Caesar's rounded that up to exactly one million dollars.

① How much extra did Caesar's give Jeff?

② Do you think they did that because they were feeling generous, to make the cheque easier to write, or to get better publicity? Write three lines to explain what you think and why.

CAESARS TAHOE PALACE
P.O. BOX 5800 • STATELINE, NEVADA 89449
FIRST NATIONAL BANK OF NEVADA
SOUTH LAKE TAHOE BRANCH
STATELINE, NEVADA
No. 19854
DATE July 16, 1981
INVOICE OR CONTRACT NO.
PAY EXACTLY 1000000
PAY EXACTLY $1,000,000.00
VOID AFTER 180 DAYS
CAESARS TAHOE PALACE
Jeffry Clinton Randolph

③ Find out from a newspaper how many American dollars are worth £1 (it's about 2). What is a million dollars worth in pounds?

What are the chances of scooping a jackpot like Jeff's? What are the odds against your winning a million?

To answer these questions you need to know something about chance, and odds.

When you spin a coin, the chance of getting heads and the chance of getting tails are exactly the same. So the chances are even. The probability of spinning heads is 50 per cent.

④ What is the percentage probability of spinning tails?

When you play a fruit machine, the chance of winning and the chance of losing are not the same. Your chance of winning is much less than 50 per cent. Your chance of winning a jackpot is much much much much less.

Fruit machines make a profit by keeping a little of the money that is fed in, and paying out the rest in occasional bursts. No machine ever pays out as much as it takes in.

So the only certainty about a fruit machine is that if you keep playing you will lose money in the long run.

⑤ Suppose a fruit machine keeps 20 per cent of the money put in. If you could play forever, how much of your money would you lose?

⑥ If Jeff Randolph's machine kept 20 per cent, what per cent did it pay back?

⑦ We know it paid back $992 018. How much profit must it have made for Caesar's Palace?

Caesar's Palace weren't at all sad when Jeff won. In fact they must have been delighted. They got lots of good publicity. And they had already made a huge pile of money out of the machine.

*The chance that someone will win the jackpot in the end is high—perhaps even 100 per cent. But the chance that **you** will win a jackpot with any one coin is very low indeed. In Jeff's case, it must have been less than one in a million.*

⑧ Your cousin has entered a competition, for which the first prize is a holiday for two. Write her a letter to explain why her chances of winning are very low.

A holiday for two

YOUR CHANCE TO WIN THE HOLIDAY OF A LIFETIME!...

Odds against

Odds and chance tell you how likely something is to happen. You can sometimes work out odds and chance by counting the number of ways in which it could happen. A spinning coin can land on one of two sides—head and tail. The two are equally likely, and so the chance of getting heads is $\frac{1}{2}$, or 0·5, or 50 per cent.

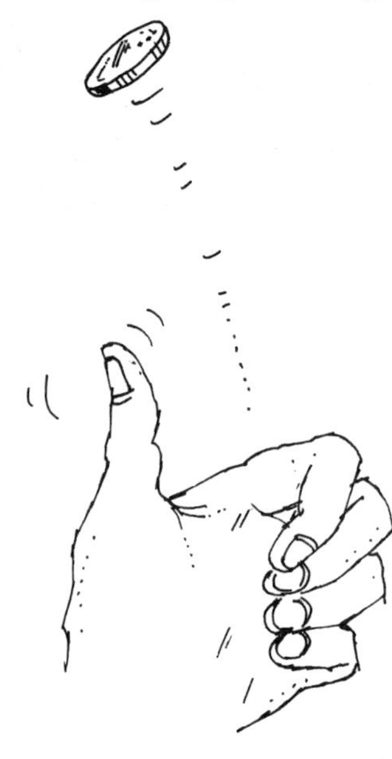

Suppose you spin a coin twice—or two coins once each. You could get two heads, or one head and one tail, or two tails. So is there a one-third chance of getting each of these results?

No, because that isn't fair. You have to allow for all possibilities for both coins:

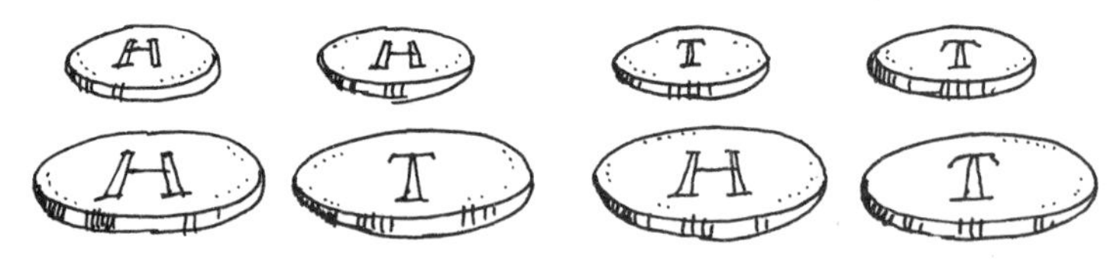

The coins could land in four different ways. Think carefully; head on the first and tail on the second is different from tail on the first and head on the second.

All these four results are equally likely. The chance of getting any one of them is one quarter. So the chance of throwing two heads is $\frac{1}{4}$, or 0·25, or 25 per cent.

This means that of all the possible ways two coins can land, one in four is with two heads on top. And it also means that if you throw two coins a million times, you will probably get two heads one quarter of those times; that is 250 000 times.

It does not mean that if you throw two coins only four times you will definitely get two heads exactly once. You may, but you may not. Chance is about probability rather than certainty.

① Write down the chance that if you throw two coins you will get (a) two tails, (b) one head and one tail.

> *The **chance** of one particular result from an event can be found by*
>
> $$\frac{\text{(Number of ways that particular result can happen)}}{\text{(Total number of possible results)}}$$
>
> *The **odds against** a result are given by this ratio:*
>
> $$\frac{\text{(Total number of possible but wrong results)}}{\text{(Total number of possible and right results)}}$$

② Write down the odds against getting two heads when you throw (a) two coins, (b) three coins.

Chance		Per cent chance	Odds against	Is it likely?
1	1	100		Absolute certainty.
$\frac{99}{100}$	0·99	99	1–99	Almost certain: 99% chance.
$\frac{9}{10}$	0·9	90	1–9	Very likely.
$\frac{1}{2}$	0·5	50	1–1	Evens, 50–50.
$\frac{2}{5}$	0·4	40	3–2	Slightly unlikely.
$\frac{1}{5}$	0·2	20	4–1	Unlikely.
$\frac{1}{10}$	0·1	10	9–1	Very unlikely.
$\frac{1}{100}$	0·01	1	99–1	One in a hundred.
$\frac{1}{1000}$	0·001	0·1	999–1	Almost impossible: one in a thousand.
0	0	0		Completely impossible. No chance.

③ On Tuesday, what is the chance that tomorrow will be Wednesday? What is the percentage chance?

There are 52 cards in a normal pack, including four aces and four kings. Imagine that you pick one card from the pack.

④ What chance do you have of getting a red card?

⑤ What is your chance of picking an ace?

⑥ What are the odds against your picking a black queen?

Dice have six sides, marked 1 to 6. Suppose you throw one die and your score is the number on the top face.

⑦ What is your chance of scoring a 6?

⑧ What are the odds against your scoring a 5?

⑨ Friday 13th is supposed to be a special day for luck. In one year, about how many 13ths would you expect to fall on a Friday? Look at a calendar; how many are there this year?

⑩ Working in pairs, throw a die 60 times. Record the number of times either of you gets a six. How many would you expect?

How many sixes do other pairs throw? Work out the average for the class, and see whether the average is closer to the expected number than your own result was. Why should it be?

Throws	Sixes
✓✓✓✓✓✓ ✓✓✓✓✓✓ ✓✓	✓✓✓

Are you psychic?

The first serious tests of psychic ability were conducted in the 1930s by Dr Joseph Banks Rhine at Duke University in North Carolina. They depended on the laws of chance.

Telepathy, or ESP (extra-sensory perception) means communication directly from mind to mind. Thought-transfer, without normal use of the senses.

No one knows whether ESP really exists, which is why Rhine wanted to test it. To do so, he invented Zener cards, which have five special symbols on them.

The complete pack contains 25 cards, 5 of each kind.

Make a pack of Zener cards, using pieces of cardboard. Draw these symbols on them with a felt-tip pen. Shuffle the pack. Take cards one at a time, face down, from the top of the pack.

The idea is to see whether a friend can identify the images on the cards, without seeing them, by reading your mind.

Ask your friend what each card is, as you turn it over. You should look hard at the symbol on the card and concentrate on it. Then your friend may have a chance of reading your mind.

- *The guesser must not see even the backs of the cards. Use a blindfold, or a screen, to prevent any cheating.*
- *Don't tell the guesser which guesses are right or wrong until you have been right through the pack.*

You may find it easier to work in groups of three. One person turns the cards, the second guesses, and the third notes the results.

Record the results in two columns; wrong guesses in one column and right guesses in the other. Check that you have 25 marks in all, and note how many right guesses there were. Then change places, shuffle the cards, and start again.

(1) There are five different cards. What is the probability of guessing a single card correctly?

(2) Guessing just at random, how many guesses can you expect to get right in one run through the pack?

You could make sure of getting five right by saying 'circle' every single time. But to show evidence of ESP you have to do better than that. You have to guess correctly more often than is probable.

If possible, have four runs each as guesser. That means 100 guesses each, which makes the chances easy to work out.

③ Out of 100 guesses, how many would you expect to be right, just by chance?

Anyone who regularly guesses more than 30 cards right out of 100 shows some telepathic ability. The odds are very high against getting more than 30 per cent right just by chance.

Psychokinesis, or PK for short, means moving things by thought alone. Some people claim they can bend spoons, forks, or keys with mental power. Others are convinced they have broken up clouds, or moved objects across a room, just by wanting these things to happen. But no one has proved it.

How strong is your will-power? You can test your PK with dice. The idea is to try and think the dice into doing what you want.

④ Pick a target number that you want—say 5. Concentrate on 5 while you are throwing. Will the 5 to stop on top. Think 5.

Throw as many dice as you like at each throw, but make sure you write down all the results.

When you have 100 results, see how many were fives.

⑤ How many fives would you expect in 100 throws?

If you get more than 25 of your target number in the 100 throws, you are doing well. But be careful not to cheat!

J B Rhine took great trouble to prevent cheating. He even built complicated machines to throw the dice fairly.

Try working in groups of three. One person tries to will the dice. The second person records the results. The third, who does the throwing, should not know what number is the target.

⑥ You could organize a collective experiment. Find out who in the class gets the best results. Then ask the two or three 'best' people to concentrate on the same number while the dice are thrown.

Or perhaps half the class could think of one number, and the other half of another number. Throw 100 dice and see which half wins. Have girls more will-power than boys?

Dracula's café

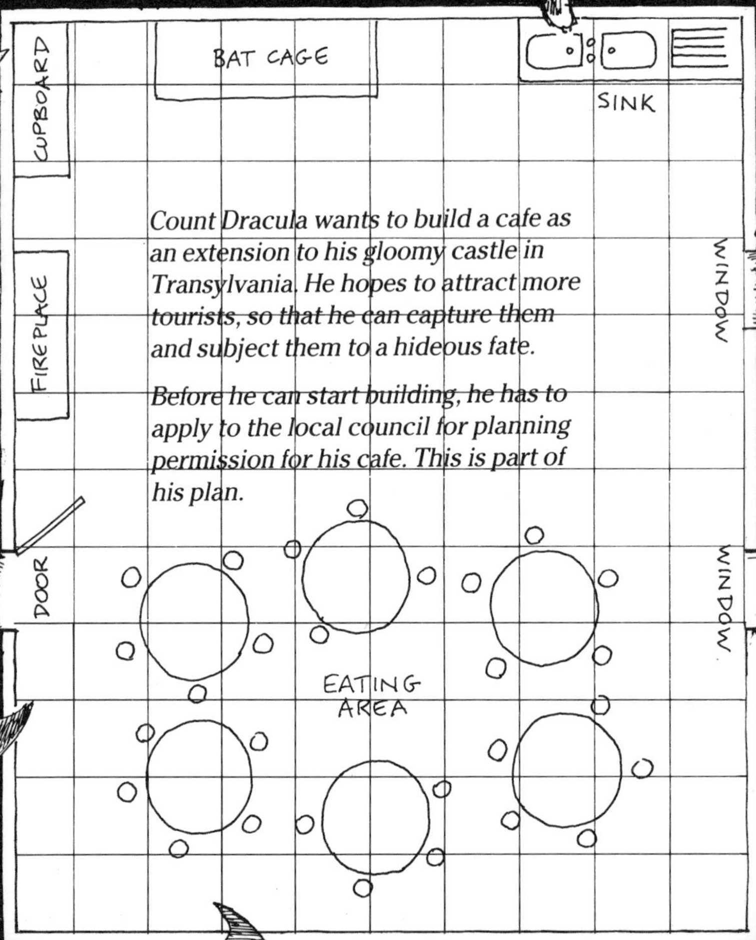

BAT CAGE

CUPBOARD

SINK

Count Dracula wants to build a cafe as an extension to his gloomy castle in Transylvania. He hopes to attract more tourists, so that he can capture them and subject them to a hideous fate.

Before he can start building, he has to apply to the local council for planning permission for his cafe. This is part of his plan.

FIREPLACE

WINDOW

DOOR

WINDOW

EATING AREA

The scale of this plan is 1 cm = 1 m; so each centimetre on the plan represents 1 metre on the ground.

① Is this the same scale as 1 : 10, 1 : 100, 1 : 1000, or 10 : 1? Which one do you think is right, and why?

② What is the length of Dracula's proposed cafe, in metres? What is the width? What is the ratio of length to width?

③ How wide will the windows be? Would a vampire bat with a wingspan of 1·5 metres be able to fly in?

One vital piece of equipment is the acid bath for disposing of the remains of victims. This has to be at least 2 metres long, to fit in a whole body. Dracula would like to put it behind the door, so that it doesn't worry new customers.

④ Is there enough space for the acid bath behind the door? Would this be convenient for the plumbing—there's a lot of acid sludge to go down the drain? Is there anywhere else in the room he could put the acid bath?

*The Count is fond of the Golden Ratio 1·618 (**see page 25**). He has found that the ratio of his height,* h, *to the height,* n, *of his navel above the ground is almost exactly the Golden Ratio.*

That is $\frac{h}{n} = 1{\cdot}618.$

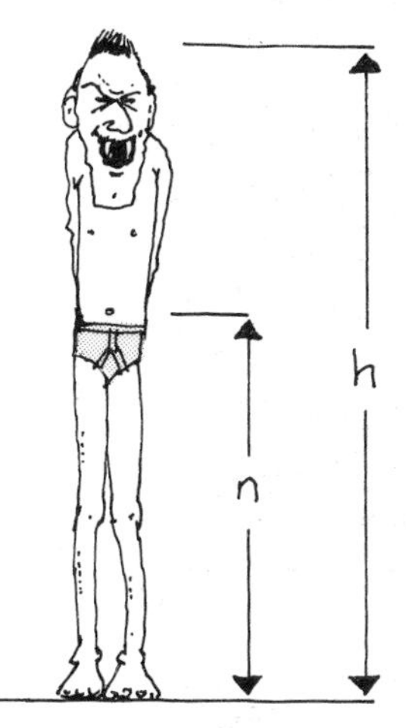

⑤ What is the ratio of your height to the height of your navel? And what is the ratio of your navel height to the rest of your height *r*?

The most popular dish on Dracula's menu is spaghetti. His own horrible recipe starts with these ingredients:

SPAGHETTI TOURISTAISE (for 4 people)

minced guts 1 kg (0·5 small tourist)
bat's blood 1·6 litre
grey matter 0·6 kg (1 brainy tourist)
pinch of herbs (but absolutely NO garlic)
spaghetti 40 m (shredded skin of 1 tourist)

⑥ When you scale up a recipe to feed more people, you have to keep the amounts of the ingredients in the same ratio as they were to start with. Write lists of ingredients for the kitchen wall to make Spaghetti Touristaise for (a) 8 people, (b) 10 people. How many brainy victims would you need in each case?

Value for money . . .

In my supermarket the baked beans come in cans of five different sizes. These are the amounts the tins contain and the prices.

150 g	225 g	450 g	580 g	840 g
15p	17p	20p	35p	40p

① Which is the cheapest can?

The fact that the smallest can is cheapest may not be a good reason to buy it. In my house we always eat more than 150 g of beans in one meal. (We don't eat baked beans at every meal, but whenever we do eat baked beans, we eat at least 300 g.)

To find out the best value for money, you have to work out a ratio. You want to know which can gives you the most beans for your money—the most beans for 1p.

② The small can gives you 150 grams for 15 pence. Work out how many grams of beans you get for one penny. Write it down as a ratio:

$$\text{(grams of beans)} : \text{(pence)}, \text{ or } \frac{g}{p}.$$

③ Work out the ratio of $\frac{\text{amount}}{\text{price}}$ for each size of can. Which gives you the most beans per penny? Which is the best value for money?

④ What would be the least I could pay at my supermarket for 300 g of beans?

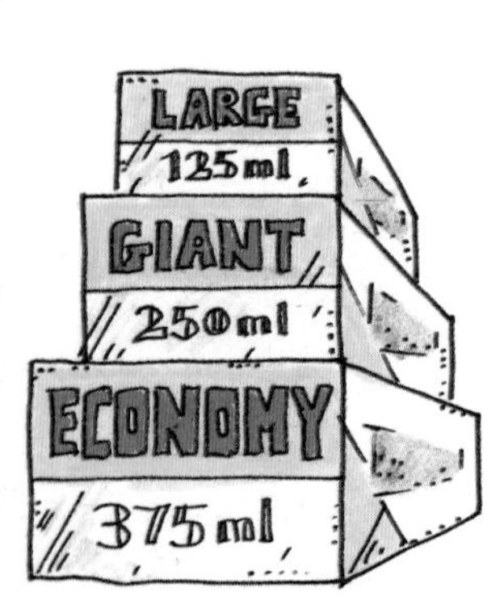

Toothpaste manufacturers have extraordinary ideas about size and price. I can never find a 'small' tube of toothpaste, nor even a 'medium'. Instead, they seem to have names like Large, Giant, Family, and Economy.

In my supermarket this week the prices of one particular brand are as follows:

Large (125 ml)	Giant (250 ml)	Economy (375 ml)
78p	132p	199p

They also have a Dispenser pack (100 ml) for 99p, and a Special Offer Twin-pack of two Large tubes taped together for 145p.

⑤ What ratio do you need to work out to find the value for money for each pack of toothpaste? Work out this ratio for each of the five packs, and write the results in a table. Which is the best buy? Is the twin-pack such good value as the shop suggests?

⑥ In general, do you think it's true that you get better value for money by buying big packets? Do your own survey in a supermarket to find out.

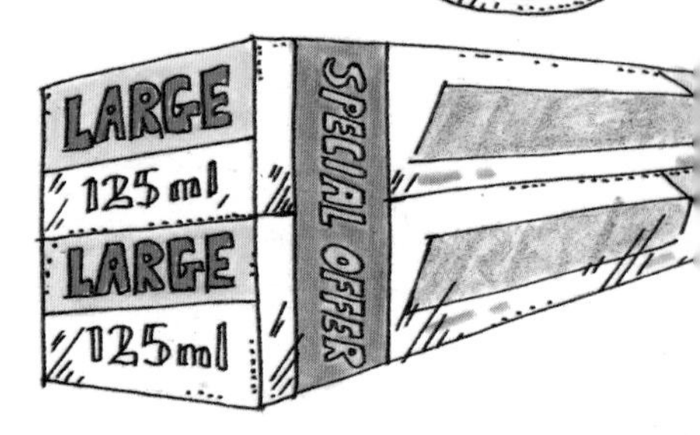

. . . and a rational holiday

You go off on a holiday to the United States. Some tricky problems turn up along the way, but you can solve them all by using ratios.

① First you have to change your money from pounds into dollars. The rate of exchange is £1 = $1·80. What will $100 cost in pounds?

② You are 135 cm high. One day in New York you find your shadow is 162 cm long. The shadow of the Empire State Building stretches 450 m up Fifth Avenue. About how high is the Empire State Building?

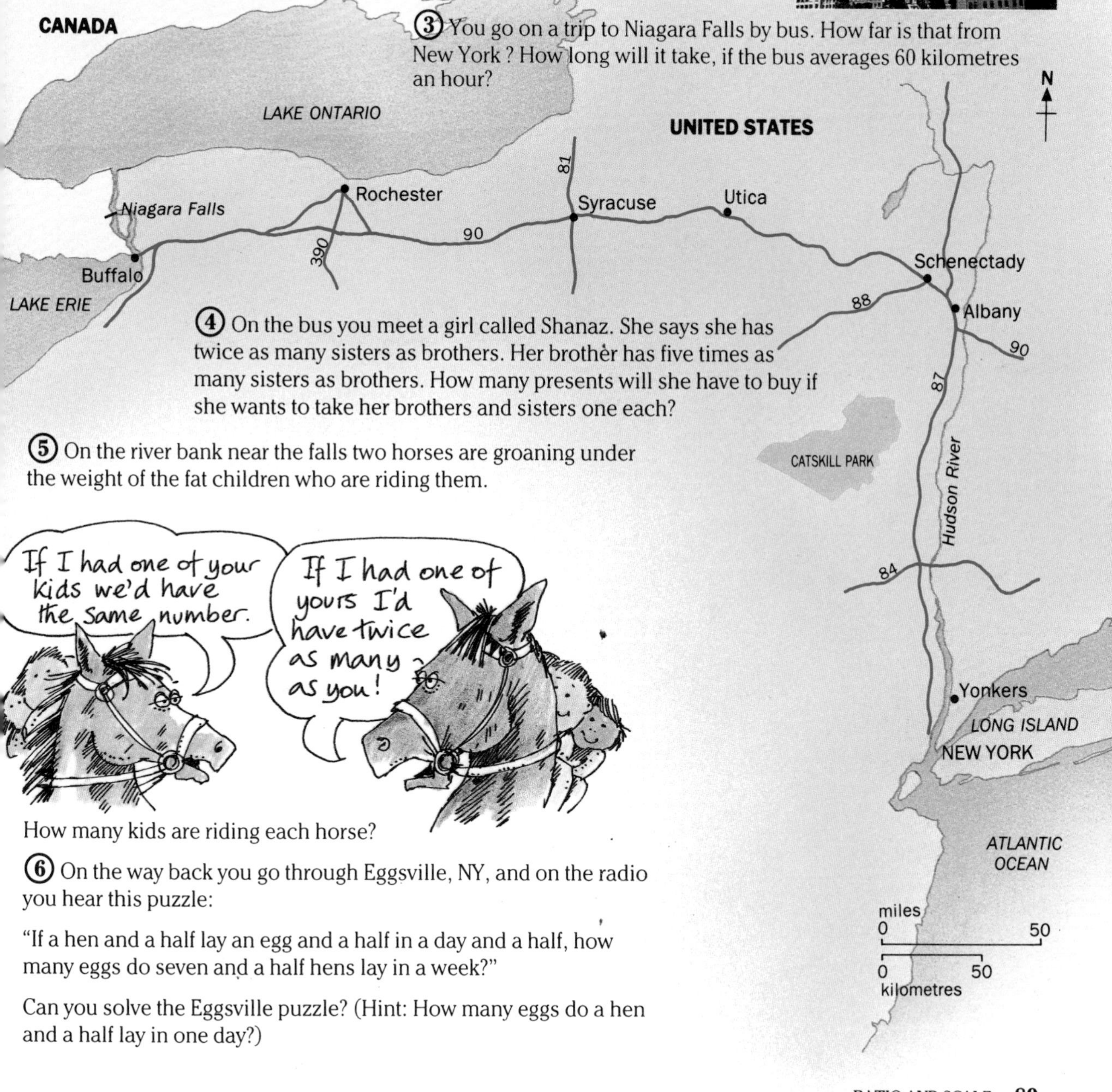

③ You go on a trip to Niagara Falls by bus. How far is that from New York? How long will it take, if the bus averages 60 kilometres an hour?

④ On the bus you meet a girl called Shanaz. She says she has twice as many sisters as brothers. Her brother has five times as many sisters as brothers. How many presents will she have to buy if she wants to take her brothers and sisters one each?

⑤ On the river bank near the falls two horses are groaning under the weight of the fat children who are riding them.

How many kids are riding each horse?

⑥ On the way back you go through Eggsville, NY, and on the radio you hear this puzzle:

"If a hen and a half lay an egg and a half in a day and a half, how many eggs do seven and a half hens lay in a week?"

Can you solve the Eggsville puzzle? (Hint: How many eggs do a hen and a half lay in one day?)

Paper and Pythagoras

When the paper-maker makes a piece of paper,
She doesn't make just any fancy shapes.
For the ratio she picks, for the long side to the width,
Is the same for every paper-size she makes.

Now A4 is just as lengthy as big A3 is wide,
And their areas are easy to predict.
For A3 is twice A4, and A2 is two times more.
So what's the length : width ratio the paper-maker picked?

The most commonly used size of typing paper is A4, which is 297 mm by 210 mm. That means the paper is 297 mm long and 210 mm wide.

① Use a calculator to work out the ratio of length to width for A4.

(Hint: Ratio $= \frac{\text{length}}{\text{width}}$)

Write it down, rounded to three decimal places.

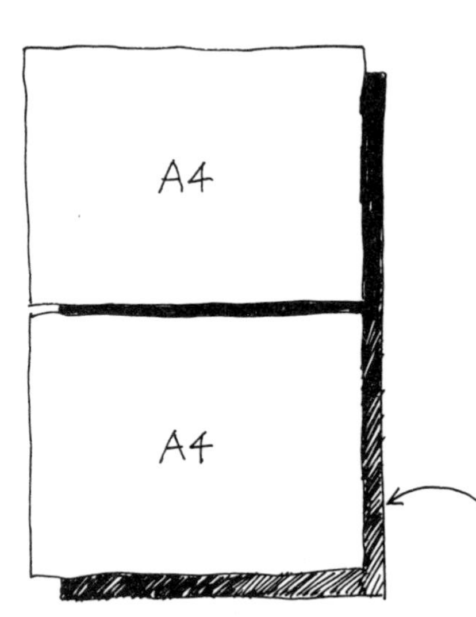

A3 is exactly twice the size of A4. Lay two pieces of A4 paper side by side and they would just cover a piece of A3 paper.

② What are the dimensions of A3 paper—i.e. length and width?

③ What is the ratio of length to width for size A3?

You can buy paper in sizes A1, A2, A3, A4, and A5. Cut any one of the first four in half across the middle, and you get two pieces of the next size down. So cut a piece of A1 in half and you get two pieces of A2.

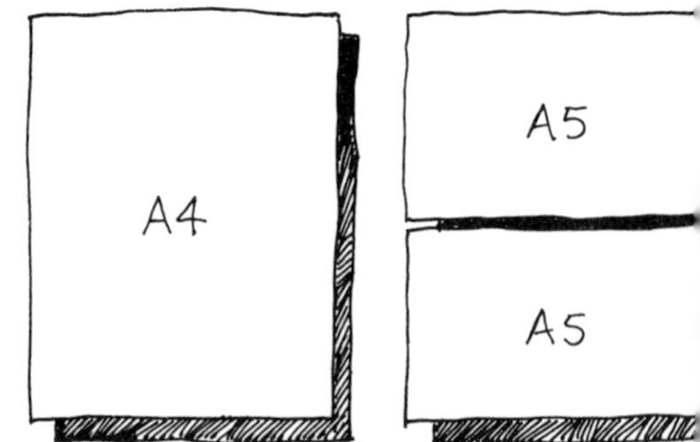

④ Work out the dimensions of A1, A2, and A5. Make a table in three columns to show length, width, and ratio of length to width, for all five sizes of paper. What do you notice about the ratios?

⑤ Work out the average value of the ratio of length to width. Multiply this number by itself and write down your answer. Can you suggest why the answer should be what it is?

Pythagoras was a brilliant Greek mathematician who ran a curious school of maths, music, and PE in southern Italy before 500 BC. His 300 students were not allowed to eat beans, wear wool, or touch white feathers.

Pythagoras's most famous theorem is: In any right-angled triangle, the square of the longest side (the hypotenuse) is equal to the sum of the squares of the short sides.

In this triangle the sides are 3 cm, 4 cm, and 5 cm long.

⑥ Work out 3 × 3, 4 × 4, and 5 × 5. Does (3 × 3) + (4 × 4) = 5 × 5?

⑦ Does the same rule apply to bigger triangles, if you keep the ratio of the sides the same? Suppose you doubled them all, to make a triangle with sides 6 cm, 8 cm, and 10 cm? What if you multiplied all the sides by 3, or 4?

Pythagoras was worried, though, by the simplest right-angled triangle of all. The short sides were 1 cm and 1 cm. How long was the other side?

Suppose its length is L.

Then we know that $L \times L = (1 \times 1) + (1 \times 1) = 2$

So $L^2 = 2$, *and* $L = \sqrt{2}$

We know that $\sqrt{2}$ *is about 1·414 (see question 5 opposite). But the Greeks had no calculators, and no decimal numbers. The only way they could talk about numbers like this was in terms of fractions—that is, as ratios of whole numbers.*

They could say one and a half was $\frac{3}{2}$. *But they could not find any ratio of whole numbers that was exactly equal to* $\sqrt{2}$, *which was not surprising, because there is no such ratio.*

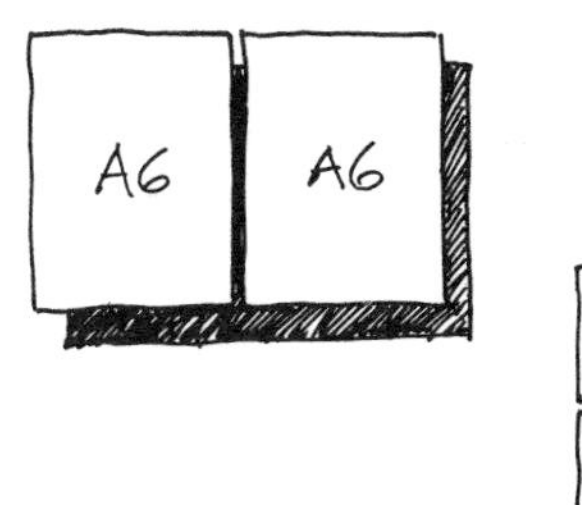

⑧ $\frac{7}{5}$ *is quite a good approximation to* $\sqrt{2}$. Use your calculator to try and find a better one. You may only use whole numbers.

Pythagoras got into terrible trouble over this triangle. His enemies said the length of the long side was 'irRATIOnal', because they could not write it as a ratio. Some people say that Pythagoras came to a sticky end because he invented irrational numbers.

⑨ Many numbers turn out to be irrational, including π, the ratio between the circumference and diameter of a circle (see page 11). Try to find a better rational approximation for π than $\frac{22}{7}$.

How to measure the Earth

In about 240 BC, a Greek mathematician called Eratosthenes wanted to find out how far it was round the world. Greek ships weren't able to travel far then; so he could not sail round and measure the distance. Instead he worked out the **circumference** *of the Earth by measuring the angle of elevation of the Sun.*

At his home in Alexandria, Eratosthenes measured the angle of elevation of the Sun at noon on midsummer day. It was 82·8°.

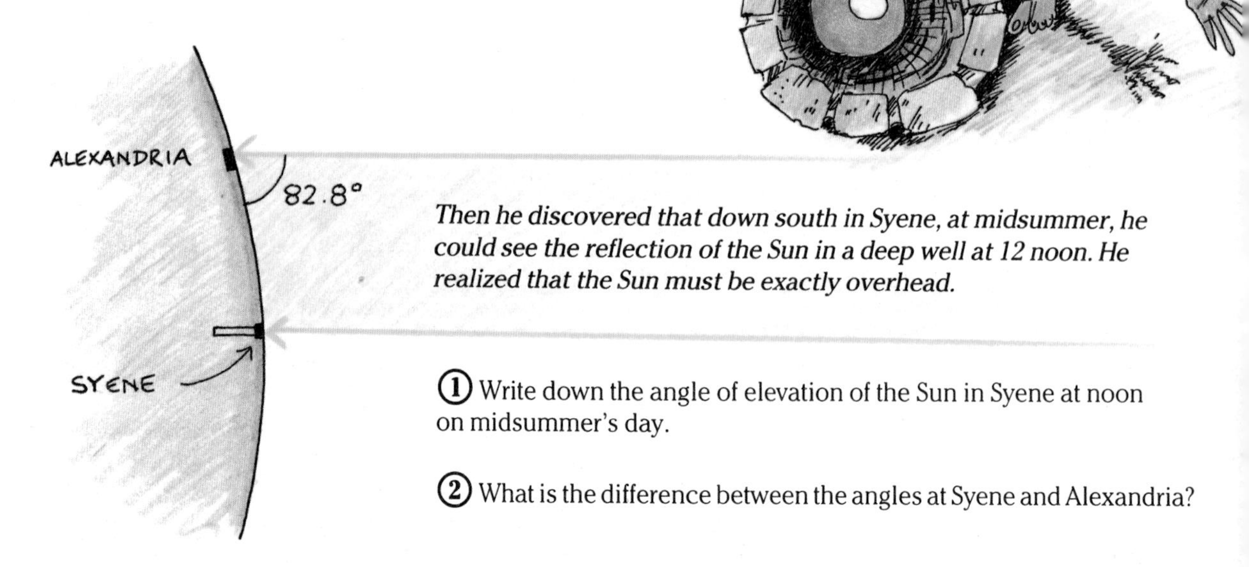

Then he discovered that down south in Syene, at midsummer, he could see the reflection of the Sun in a deep well at 12 noon. He realized that the Sun must be exactly overhead.

(1) Write down the angle of elevation of the Sun in Syene at noon on midsummer's day.

(2) What is the difference between the angles at Syene and Alexandria?

Eratosthenes believed the Earth to be round. He argued that the Sun was very far away; so all the rays of light coming from it must be parallel.

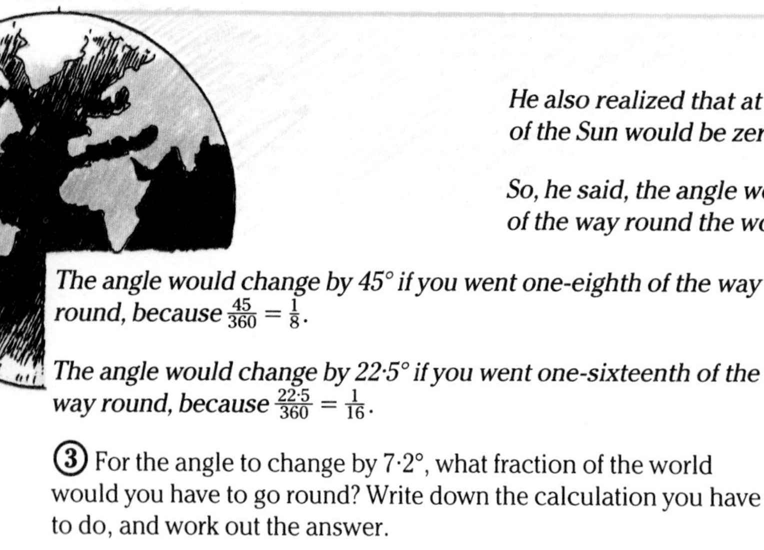

He also realized that at the top of the world the angle of elevation of the Sun would be zero.

So, he said, the angle would change by 90° if you went one-quarter of the way round the world, from Syene to the top. $\frac{90}{360} = \frac{1}{4}$.

The angle would change by 45° if you went one-eighth of the way round, because $\frac{45}{360} = \frac{1}{8}$.

The angle would change by 22·5° if you went one-sixteenth of the way round, because $\frac{22 \cdot 5}{360} = \frac{1}{16}$.

(3) For the angle to change by 7·2°, what fraction of the world would you have to go round? Write down the calculation you have to do, and work out the answer.

Eratosthenes knew that the distance from Syene to Alexandria was 5000 stades, as measured by the army. Soldiers were ordered to count their paces when they marched.

He reckoned that the distance from Syene to Alexandria must be $\frac{1}{50}$ of the circumference—the whole way round the Earth. He also knew the distance was 5000 stades, he could write an equation.

$$\frac{1}{50} \times \text{circumference} = 5000 \text{ stades}$$
$$\text{so circumference} = 50 \times 5000 \text{ stades}$$
$$= 250\,000 \text{ stades}$$

He was close. So were these mathematicians on 20 January 1988.

Angela and her class at Anderson High School in Lerwick, Shetland, measured the angle of elevation of the sun at noon, and got 9·7°.

Lerwick is 700 km due north of Leeds.

At Shakespeare Middle School in Leeds, Yorkshire, Sam and Craig measured the angle as 15·9° at the same time on the same day.

④ Write down the difference between the angles of elevation at Leeds and Lerwick.

⑤ Use this difference to work out the circumference of the Earth in kilometres.

⑥ Look in an atlas in your library to see what the real value is.

⑦ If Eratosthenes was right, how long was a stade?

⑧ How else could the circumference of the Earth be measured? Write down some of your ideas. See if you can find out how else it has been done.

Polygon crazy

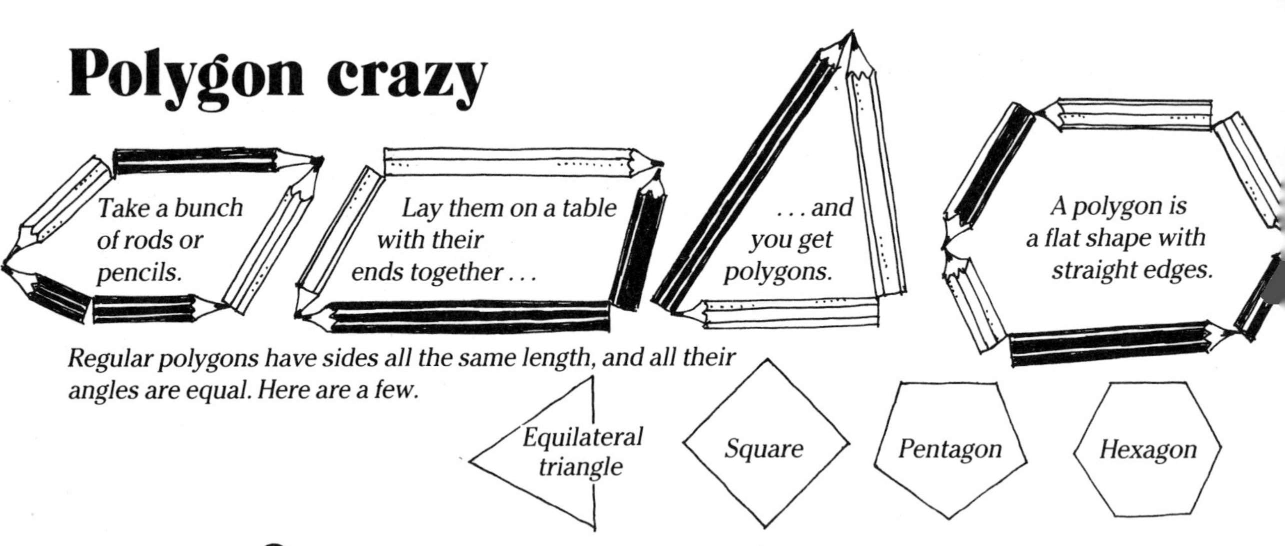

Regular polygons have sides all the same length, and all their angles are equal. Here are a few.

① Draw a square by using the end of your ruler twice. Measure the length of each side and the size of each angle.

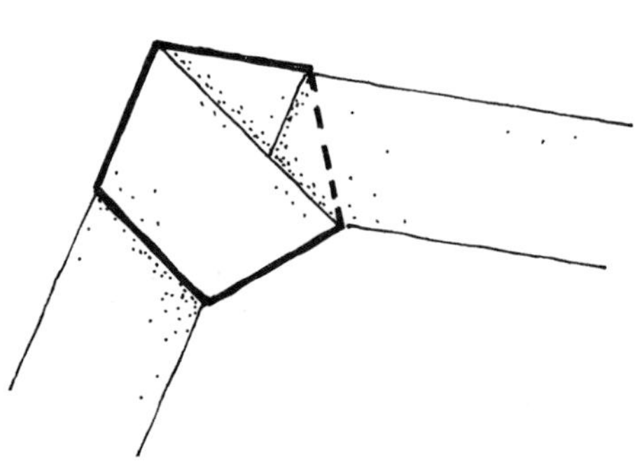

② Make a pentagon by tying a knot in a strip of paper that is at least 20 cm long, and the same width—about 3 cm or 4 cm—all along its length. You can make this strip by first folding a piece of newspaper 3 cm or 4 cm from one edge, and then carefully cutting or tearing along the fold.

Tie the strip gently into a simple knot. Carefully pressed flat, the knot forms a regular pentagon. Draw round it to get the shape on paper. Measure the angles. If they aren't quite equal, work out the average angle.

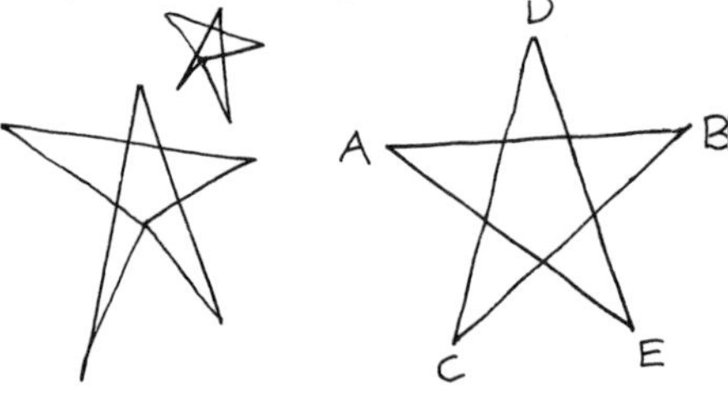

A five-pointed star is sometimes called a pentagram. Pythagoras used a pentagram as a sign for good health.

③ Practise drawing five-pointed stars. Go from A to B to C to D to E, trying to keep each line the same length. The angles in the points should be about 36°. Can you see how your pentagon knot could help?

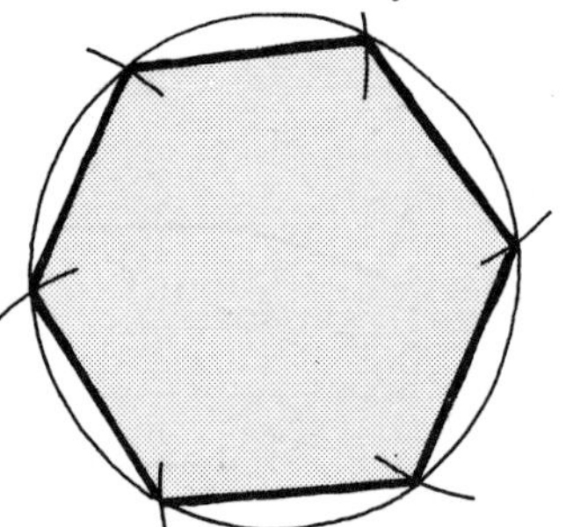

④ Draw a hexagon by using a pair of compasses. First draw a circle about 5 cm in diameter. Keep your compasses the same size, put the point anywhere on the first circle, and draw an arc (part of a circle) that crosses the first circle.

Then put the point where the lines cross, and draw another arc. Do this again, and carry on until you have drawn six arcs round the circle. Use a ruler to connect each crossing point to the next. Behold! A regular hexagon. Measure and write down the angles.

⑤ A diagonal is a straight line that joins two corners of a polygon but does not form an edge. You can't draw any diagonals in a triangle. You can draw two in a square.

How many diagonals can you draw in a pentagon? What about a hexagon? Can you predict how many you could draw in a shape with seven sides? Can you write down a general rule?

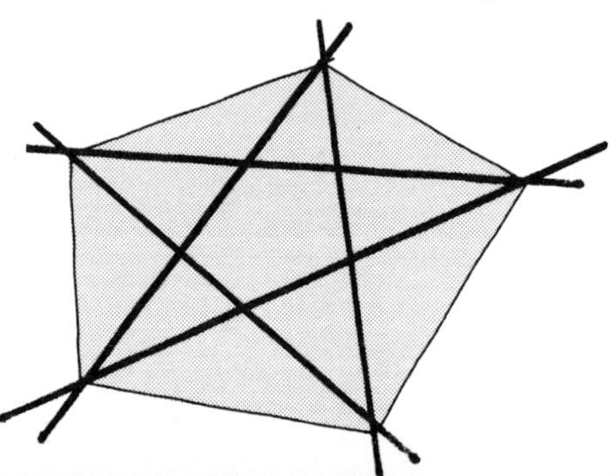

Polygon Pete

Polygon Pete is an angular cowboy. His job is to check the fence that keeps the steers in and the rustlers out. To do that he rides all the way round the corral, and back to where he is now.

① How many degrees has Pete turned through?

Try moving a rubber, or pencil, once around any object on the table. You have to turn the rubber or pencil through 360°. The shape of the corral does not matter; whether it has three sides, or thirty-three, the answer is the same.

Suppose the corral is an equilateral triangle. Start Pete in the middle of one side.

At the first corner he has to turn through 120°. Same at the second, and the third.

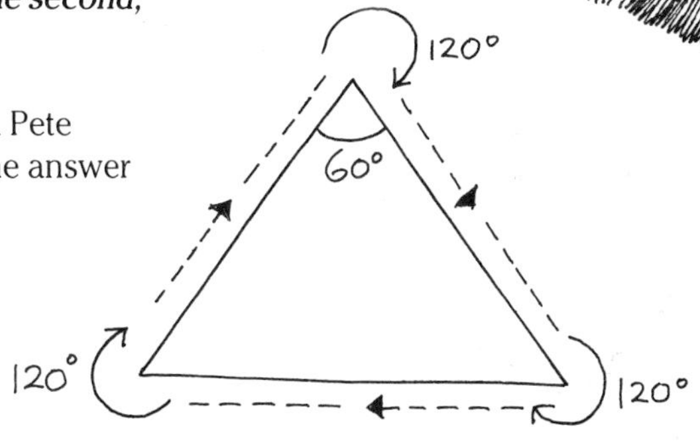

② Add the angles to find the total amount through which Pete would turn if he toured right around the triangle. Would the answer be the same for any triangle?

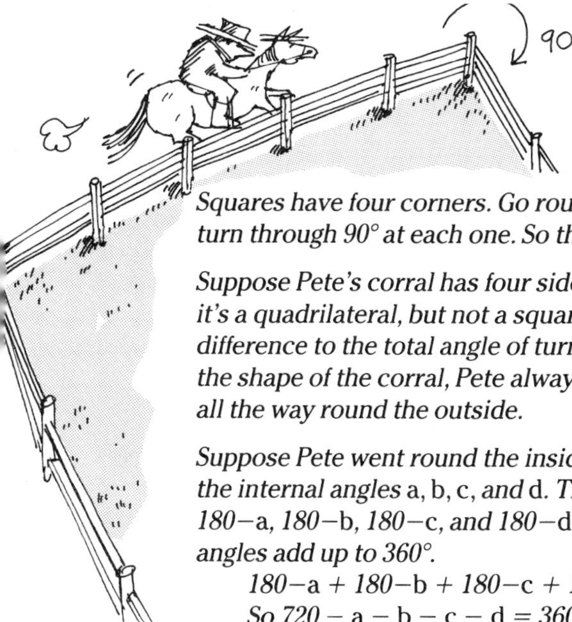

Squares have four corners. Go round the outside, and you have to turn through 90° at each one. So the total turn is 4 × 90° = 360°.

Suppose Pete's corral has four sides, but all of different lengths; so it's a quadrilateral, but not a square. That doesn't make any difference to the total angle of turn round the outside. Whatever the shape of the corral, Pete always has to turn through 360° to get all the way round the outside.

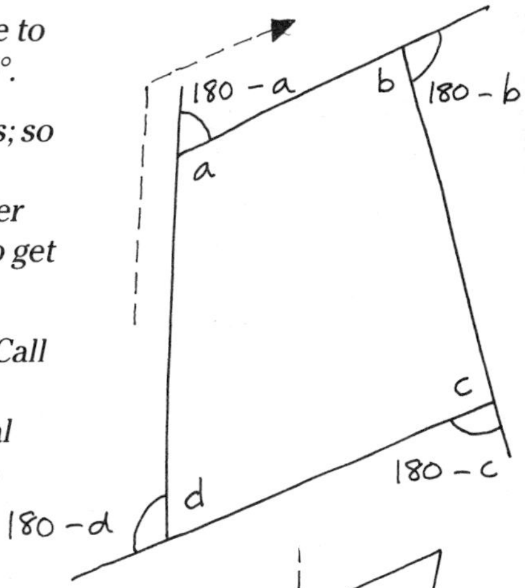

Suppose Pete went round the inside of a quadrilateral corral? Call the internal angles a, b, c, *and* d. *Then the external angles are* 180−a, 180−b, 180−c, *and* 180−d. *And we know these external angles add up to 360°.*

$$180-a + 180-b + 180-c + 180-d = 360$$
$$\text{So } 720 - a - b - c - d = 360$$
$$\text{So } a + b + c + d = 360$$

Aha! So the angles inside a quadrilateral add up to 360°, regardless of its shape.

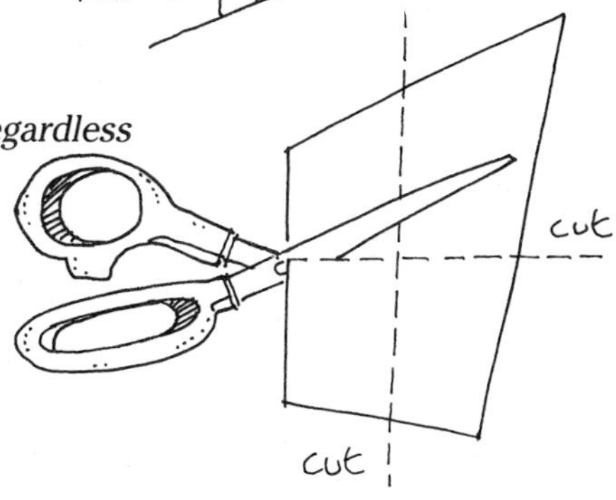

③ Don't take this on trust; check it. Draw a quadrilateral on a piece of paper. Cut it out, and then cut it across twice, as suggested here. Try fitting the cut-out corners together at one point. If they fit, then they add up to 360°.

See you later, tessellator...

Many bathroom walls are covered with tiles. The tiles are meant to look attractive and keep water away from the plaster underneath. To keep out the water, the tiles must fit together exactly, leaving no gaps. Tiles that fit together and cover a flat surface without leaving gaps are said to ***tessellate****.*

To make tiles look attractive, tile manufacturers often colour them and put pictures on them. But they nearly always use the same shape; most bathroom tiles are boring old squares.

Tiles don't have to be square; they can be other shapes.

① See whether you can fit 20 identical triangles together, leaving no gaps. An easy way to make 20 'tiles' the same is to draw the shape on the front of a newspaper, and then carefully cut the same shape through all the pages.

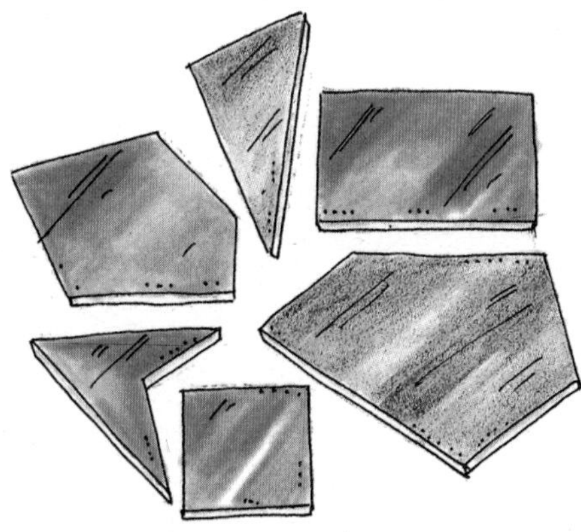

You can try to fit together tiles of any old shapes, but they won't always tessellate. If the shapes are to tessellate, when you bring together the corners at a point, the angles round that point must add up to 360°.

② Explain clearly what will happen if the angles add up to more or less than 360°.

The internal angles of any quadrilateral add up to 360°. You might suspect that if you had many quadrilaterals of the same shape they would fit together somehow. In other words, you might guess that any quadrilateral would tessellate.

③ On a newspaper, draw a quadrilateral—any shape with four straight sides. Cut out a dozen or more 'tiles', all exactly the same size and shape. Can you work out how to fit them together without leaving gaps? You may find it easier if you colour half of them one colour and the other half a different colour.

Can you do it with a shape like this?

?

What about this sort of quadrilateral?

④ What other shapes can you find that will tessellate? Will pentagons? Or hexagons? Or pentagrams?

⑤ Footballs are made from pentagons and hexagons fitted together. Find out how many of each shape there are. Why do you think two different shapes are used?

⑥ M S Escher was a graphic artist who made many marvellous tessellations. This picture of fish and birds is like one of his. Either copy this picture, or find one of Escher's designs in the library, and copy that. Colour your copy in a way that will bring out the pattern as strongly as possible.

⑦ Can you colour the same picture in two different ways, to bring out different ideas in the tessellation?

⑧ Imagine you are a designer at Trendy Bathrooms and Kitchens plc. Design a tile, or a set of tiles, that would be attractive and more interesting than boring old squares.

Fungus

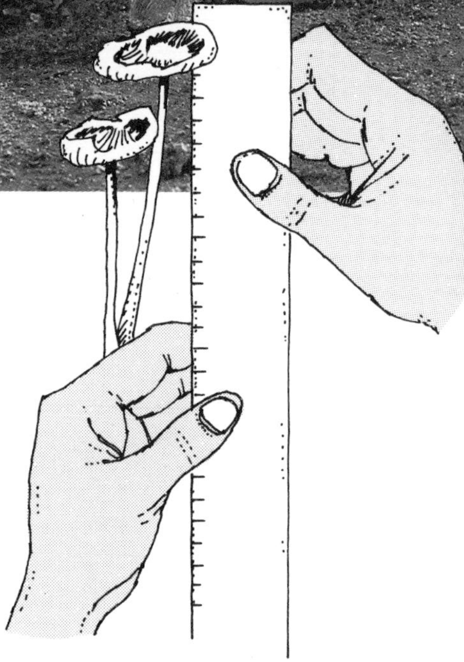

I found this fungus growing on an elm log in a shed.

Fungi are like tiny plants that grow underground, and inside wood. They live by eating the wood. That is how wood rots.

Most of the fungus is inside the wood. The bits you can see are called 'fruiting bodies'. They appear only once a year. In this photograph they are the same size as they were on the log.

① Draw a rough sketch of the front row of fruiting bodies. Ignore the ones behind, but label those whose stalks you can see down to the bottom, so that you will be able to identify them.

② Measure and record the height of each one in the front row, to the nearest centimetre. Label the heights according to your sketch, and write them down in order of increasing size. Call this list A.

One way to say roughly how tall the fruiting bodies are is to pick one of them to represent the group.

③ Can you suggest which one is in the middle of the height range? If you pointed at it, would it be fair to say that most of them are about that high?

Another way to give a general idea of the height is to work out an ***average*** *height. There are various different kinds of averages; the median, the mode, and the mean average.*

To find the median, look at your list A, the order of increasing size. The median is the one in the middle. So if there are seven in your list, the median is the fourth. The median height is therefore the height of the one in the middle of this list.

④ What is the median height of the fruiting bodies?

less than 3cm	3–4cm	4–5cm	5–6cm	6–7cm	7–8cm	more than 8cm

The mode is the most popular height. Again look at your list A, and see which height is most common. For example, if four stalks are 10 cm high then 10 cm is probably the mode.

⑤ Write down the modal height of the fruiting bodies.

To work out a mean average height you add up all the heights you have measured, and divide by the number of heights you measured.

Suppose the heights of three children are 130 cm, 135 cm, and 146 cm. What's their average height?

Total height = 130 cm + 135 cm + 146 cm
= 411 cm

Average height = $\frac{411}{3}$ *cm*
= 137 cm

None of the children is exactly the mean average height, but it gives a rough idea of how tall they all are.

⑥ Work out the total height of the fungus fruiting bodies that you measured. Divide by the number of measurements to find the mean average height.

⑦ Now compare the median, the mode, and the mean heights. Are they all the same? Write five lines to say which sort of average you think is best, and why.

⑧ Find out the total number of people—adults and children—who live in the house of each person in your class. Work out the mean. Draw a bar chart of the numbers. What is the mode?

In this country as a whole the mean average number is 2·5 people living in each house. Is your average figure greater than this? Can you suggest reasons why this should be so?

Pies

Advertisements often use numbers to try to persuade you to buy something. "Three out of four people use Glucko soap!" they say, hoping that you will rush out and buy some. Numbers like this are called statistics. Sometimes the advertisers use statistics in ways that aren't quite fair.

The Great American Pie Company did a survey to find out how popular each of their pies was. They asked 36 children, "What sort of pie do you like best?" These were the answers:

15 said apple.
7 said cherry.
6 said strawberry.
4 said blueberry.
1 said chicken.
The rest said they didn't like pies.

(1) Draw a pie chart to show what fraction of children liked each kind of pie.

(Hint: Draw a circle for the pie. Let 10° represent one child, so that 360° represents 36 children.)

The company then launched an advertising campaign. They used statistics on every poster. Some of these posters are truthful and fair. Some are not.

② For each poster, write down whether you think it is (a) truthful and (b) fair, and why. If you haven't enough information to tell whether it's true, say so!

Imagine that you are the manager of the Strawberry Pie Division of the Great American Pie Company. Your job is to increase the sales of strawberry pies.

③ Draw your own poster to try to persuade people that strawberry is best. Use some statistics on your poster.

④ Look out for three real advertisements that use statistics, either on television, or in newspapers or magazines. For each one write two lines about whether you think it is fair or not, and why.

⑤ Do your own survey in class to find out what sort of pies people like. Draw your own pie chart of the results. How would you use your statistics to persuade people to eat more of the least popular pie?

Road accidents

Every year, hundreds of children aged between 10 and 14 are killed or seriously injured in road accidents.

Everyone would like to cut down the number of accidents. The best way to do that is to find out what causes them. First, think about when in the year you might expect most accidents.

① (a) Write down which month or months of the year you think might be the most dangerous for children on the roads.

(b) Do you think the same months would be most dangerous for both cyclists and pedestrians?

(c) Do you think boys or girls are more at risk?

To find out whether your ideas are correct, you need information. The police collect information at all accidents where people are killed or seriously injured. They pass it on to the Department of Transport, who provide statistics about all kinds of accidents.

The statistics are just a mass of numbers, but if you look at them in the right way they can tell you a lot about what the causes of the accidents might be.

CHILDREN AGED 10 TO 14 KILLED OR SERIOUSLY INJURED IN EACH MONTH OF ONE YEAR												
	Jan	Feb	Mar	Apr	May	Jun	Jul	Aug	Sep	Oct	Nov	Dec
On foot												
Boys	152	165	135	152	128	114	107	84	138	154	160	139
Girls	106	81	93	88	93	86	81	55	97	103	88	98
Cycling												
Boys	62	28	43	72	98	90	95	110	87	91	58	47
Girls	10	2	4	17	9	19	14	23	24	14	6	9

One good way to make sense of statistics is to show them in the form of a chart. Charts help you look for patterns in the numbers.

② Make two bar charts, one for pedestrians and one for cyclists. Show both boys and girls on each. Put the months along the bottom, and the numbers up the side.

(a) Which are the most dangerous months? Can you explain why they are different for pedestrians and for cyclists?

(b) In general, are boys or girls more at risk? Is there a bigger difference for cyclists than for pedestrians? Could you show this by working out what percentage of victims are boys?

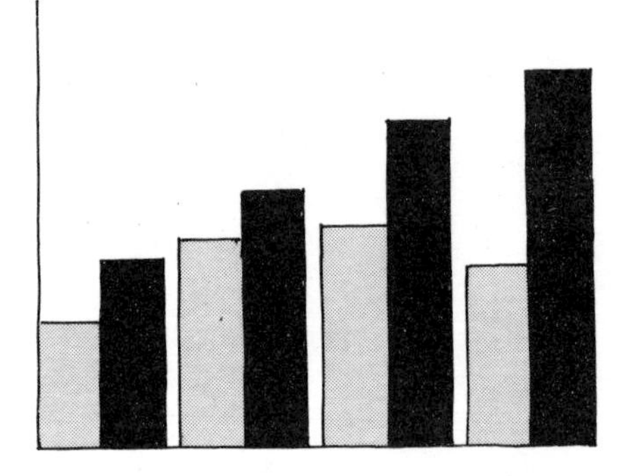

③ What times of day would you expect to be the most dangerous? (Hint: When are most children out on the roads?) Will this be different for boys and for girls?

This table shows the number of children killed or seriously injured during each hour of the day, from 7 a.m. to 8 p.m.

TIME OF DAY (7 means 7 a.m. to 7.59 a.m., 13 means 1 p.m. to 1.59 p.m. etc.)															
	7	8	9	10	11	12	13	14	15	16	17	18	19	20	21
Boys	60	230	39	55	96	179	130	133	349	359	267	201	196	115	61
Girls	14	151	25	20	30	76	89	53	177	218	109	90	76	32	34

④ (a) Make a bar chart to show this information, with the times along the bottom, and the numbers up the side. Show the numbers for boys and girls on the same chart.

(b) Can you explain why there are dangerous periods both in the morning and in the afternoon? What is happening then?

(c) Is there much difference in the patterns for boys and for girls?

Suppose you were advising the minister in charge of road safety. Your job is to cut down the number of accidents. First you must choose your main target group; which children are most at risk?

Then you need to organize a campaign aimed at those children, to try to cut down the risk. Think about how you could go about that.

⑤ Get together with a small group in your class, and work out what advice to give the minister. Then **either** write down a plan to reduce the accident rate . . .
or ask for permission to go and talk to the people in another class about the problems of road safety. Use statistics to persuade the group most at risk to be more careful.

⑥ Try to make people in your school more aware of road safety. Design a poster to change people's habits. Remember what the advertisers do? They use statistics to try and persuade people. You can use statistics too—to save people's lives.

Tangram

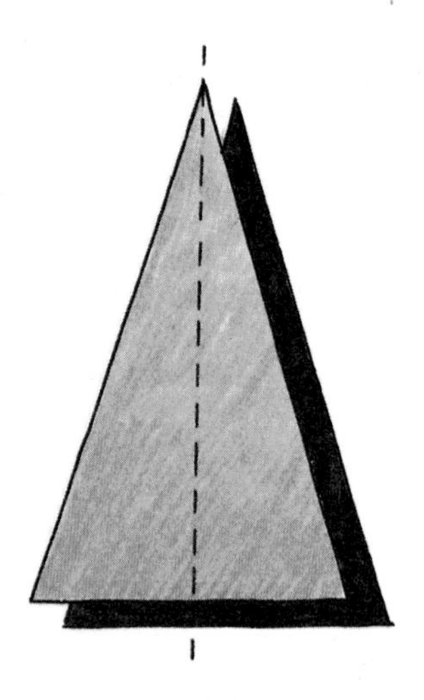

This triangle has a line of symmetry. That means four things:

- *Cut it down the dotted line, and one piece will fit exactly on top of the other piece, after you turn it over.*
- *Fold it down the dotted line, and the two halves will fit together all the way round the edge.*
- *Measure at right angles to the dotted line, and wherever you measure, the distance to an edge on one side will be matched by the same distance to another edge on the other side.*
- *Put the edge of a mirror along the dotted line, and when you look from the right direction the triangle will look the same as it did without the mirror.*

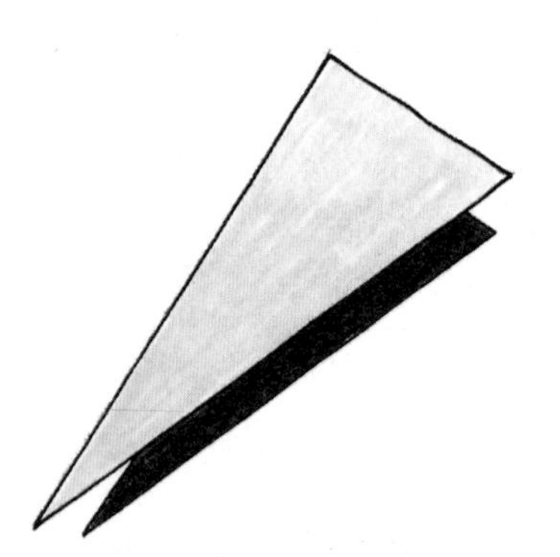

The triangle above is isosceles; that is, two of its sides are the same length. It has one line of symmetry. Some triangles have more. Some have less.

① An equilateral triangle has all three sides the same length. Draw an equilateral triangle, and draw in all the lines of symmetry you can find. How many are there?

② What is the name of a triangle with sides all different in length? How many lines of symmetry does it have?

Any shape with four sides is called a quadrilateral. The most symmetrical quadrilateral is the square.

③ Draw a square. Dot in all the lines of symmetry. Write down how many there are.

The sides of this parallelogram are not all the same length. Can you find a line of symmetry? You shouldn't, because there aren't any.

But it does have a point of symmetry. That means two things:

- *Place a ruler* **in any direction** *against the point of symmetry. Every edge it touches on one side of the point will be matched by an edge an equal distance away on the other side of the point.*
- *Rotate the shape around the point, and before it turns through 360° it will reach at least one other position that is the same as its original position.*

④ Can you find a quadrilateral that has a point and only two lines of symmetry?

⑤ What sorts of triangles have points of symmetry? Cut a triangle from cardboard, and label the corners A, B, and C. Draw round it, and label the corresponding corners of the drawn shape A, B, and C.

Then rotate the cardboard shape (without turning it over) to see if it fits on the drawn shape in any other than the original position.

⑥ A square has four lines and one centre of symmetry. Can you find any shapes that have five or more lines and a centre of symmetry? What about a pentagon, or a pentagram (see page 94)?

⑦ Is a chess-board as symmetrical as it looks? Write down what symmetry it has.

This set of seven shapes is known in the West as the ***Tangram****. It was invented hundreds of years ago by the Chinese, who call them the 'clever pieces'.*

⑧ Make a list of the shapes and symmetries of the seven pieces of the Tangram.

⑨ Copy or trace the tangram shapes, and cut your own pieces out of paper or cardboard.

What designs can you make, using all seven pieces, with (a) a point of symmetry, (b) one or more lines of symmetry, (c) no symmetry?

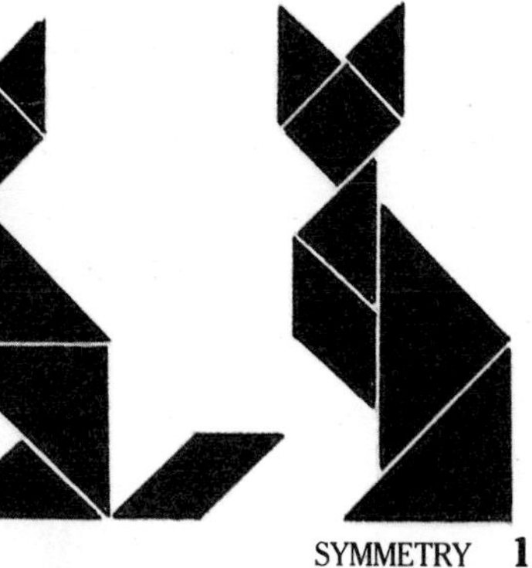

Mirror images

Letters and numbers are full of symmetry. Look for it, and it jumps out all over the place. Take the letters **A**, **K**, *and* **Z**.

The letter **A** *has a vertical line of symmetry, down the middle; the left and right sides are the same.*

K *has a horizontal line of symmetry across the middle; the top half is the same as the bottom.*

Z *has no line of symmetry; try and draw one, and it won't work. But* **Z** *does have a point of symmetry. Rotate it through 180° about the point, and its new position exactly matches its original position.* **Z** *is said to have rotational symmetry.*

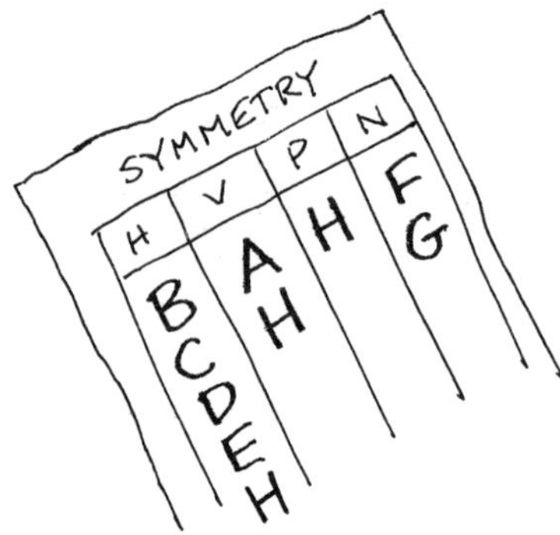

① Make a 'symmetry table' with four columns, labelled *Horizontal, Vertical, Point,* and *None.*

List the capital letters of the alphabet in the appropriate columns. **A**, for example, goes in the second column because it has vertical symmetry. Some letters belong in more than one column. Then do the same with the numerals 0 to 9.

Letters with a vertical line of symmetry look the same back to front—i.e. in a mirror.

Letters with a horizontal line of symmetry look the same upside-down—i.e. in a mirror placed along the top or the bottom of them.

② Which of these statements is true?
(a) Any letter with both horizontal and vertical l
has rotational (point) symmetry too.
(b) Letters with rotational symmetry have no ho
lines of symmetry.
(c) A letter with rotational symmetry may have n
or it may have two or more, but it cannot have o
(d) Letters with rotational symmetry never look

② Which of these statements is true?

(a) Any letter with both horizontal and vertical lines of symmetry has rotational (point) symmetry too.

(b) Letters with rotational symmetry have no horizontal or vertical lines of symmetry.

(c) A letter with rotational symmetry may have no lines of symmetry, or it may have two or more, but it cannot have one line of symmetry.

(d) Letters with rotational symmetry never look right in a mirror.

When you've sorted the letters, you can make symmetrical words. The word **SWIMS** *has rotational symmetry. Back to front it's a mess, but rotate it through 180° and you match its original position.*

Look at this family:

MUM *has a line of symmetry down the middle. Look in a mirror held along that line and* **MUM** *would look the same.*

③ Three of these words have different kinds of symmetry. The fourth has none. Work out and write down which has what. If possible, check your conclusions by using a mirror.

④ Think of some words that have symmetry. Can you find words of three letters, four letters, and five letters, for each kind of symmetry? Can you write down a sentence that would look the same in a mirror?

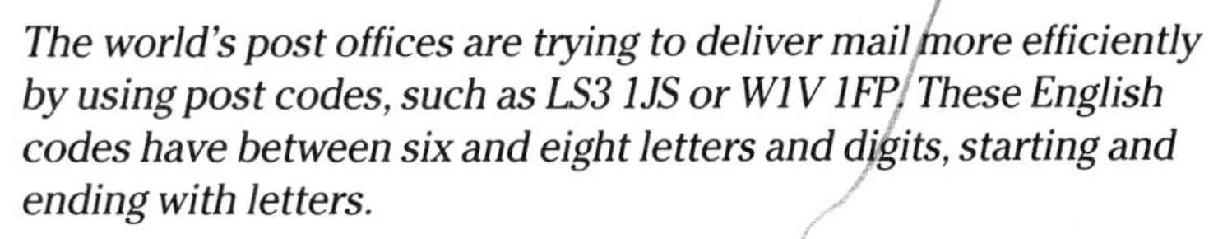

The world's post offices are trying to deliver mail more efficiently by using post codes, such as LS3 1JS or W1V 1FP. These English codes have between six and eight letters and digits, starting and ending with letters.

Canadian post codes have six letters and digits, which alternate e.g. M4T 1N5 and T6G 2G2.

⑤ Write down three English post codes, each with a different kind of symmetry. Do the same with Canadian post codes. Can you do it without using O, I, 0, or 1?

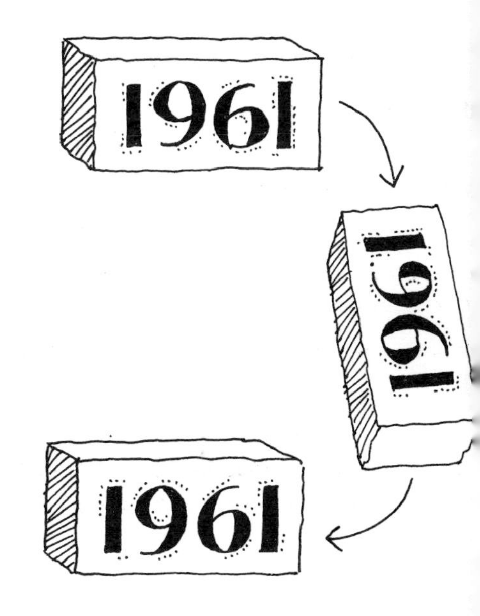

1961 was a good year. President John F Kennedy announced plans to land on the Moon. The Beatles began singing in The Cavern. And most exciting of all, 1961 marked the beginning of the Swinging Sixties by having rotational symmetry.

⑥ Which was the last year before 1961 that had rotational symmetry? Which will be the next?

If you include the day and month, then the tenth of August 2001 (10 8 01) has rotational symmetry.

⑦ What other symmetry does it have? What symmetrical dates were there in 1961, apart from 1 September? Are there any this year?

Yreka Bakery

Three hundred miles north of San Francisco, and 25 miles north of Weed, lies the Californian town of Yreka. Thirty thousand people live there, logging, fishing, and welcoming tourists. In the Old Town there are lovely shops, including some that sell bread.

One bread shop could be world-famous among lovers of palindromes.

Why? Because the words YREKA BAKERY have the same letters going from right to left as they do from left to right.

In other words, if you ignore the space, YREKA BAKERY reads the same forwards as backwards. That's what a palindrome is. Alas! None of the bread shops is called YREKA BAKERY.

① Write down which four words in this sentence are palindromes:

Madam Hannah uses the rotavator to get her green garden level.

② Suppose that in Arizona there was a town called Yrecor. Can you imagine what shop there would have a palindromic name?

Palindromes sometimes have symmetry, but not often. 1991 has no symmetry. Upside down, it becomes 1661. Back to front, in a mirror, it looks a mess.

MADAM has no symmetry either. But both 1991 and MADAM are the same from right to left and from left to right.

③ Write down a palindromic number with two digits, one with three digits, and one with four. No digit should appear more than twice in any number; so 2222 doesn't count.

④ What's the biggest palindromic number you can think of, if no digit appears in it more than twice?

You can put your calculator to work making palindromic numbers. All you need is a few tricks to start with.

5 Multiply 11 by 11 and you get 121, a palindromic number. Will you always get a palindromic answer if you multiply a number by 11? Or is it better to multiply by other multiples of 11, such as 55 or 77?

What if you multiply numbers by 101? Or 111?

Investigate ways of getting palindromes, and write down some rules to explain how to do it.

6 Here's something to baffle your friends. Think of a number—any number—though it's harder to work with big numbers. Write it down. Then write it backwards, and add the two numbers together. Write down the total and then write the total backwards. Add together these two numbers. Go on like this, and in the end you will always get a palindrome. . . . Or will you?

3519
+ 9153
= 12 672
+ 27 621
= 40 293
+ 39 204
= 79 497

7 Finding word palindromes is difficult. Here are a few:

ANNA BOB DAD DEED DID LIL MUM NOON OXO POP REPAPER SIS TOOT

How many more can you find?

Making up sentences that work as complete palindromes is even harder. When Adam first met Eve in the Garden of Eden, he is supposed to have said "MADAM, I'M ADAM."

A French engineer called Ferdinand de Lesseps first had the idea of sawing through Panama to cut America in half. Alas, he did not quite finish the job, but his epitaph might have been

A MAN, A PLAN, A CANAL, PANAMA!

Imagine a very fat man who likes fish but eats too much. His doctor tries to make him fast—i.e. go on a diet. He refuses:

DOC, NOTE, I DISSENT. A FAST NEVER PREVENTS A FATNESS. I DIET ON COD.

8 Write a story that includes five or more palindromes. Use any of those on this page, or your own. Here are a few more that may help:

ADAM I'M ADA PAM'S MAP GOD'S DOG RATS LIVE ON NO EVIL STAR

And don't forget YREKA BAKERY!

. . . and a partridge in a pear tree

On the twelfth day of Christmas, my true love sent to me

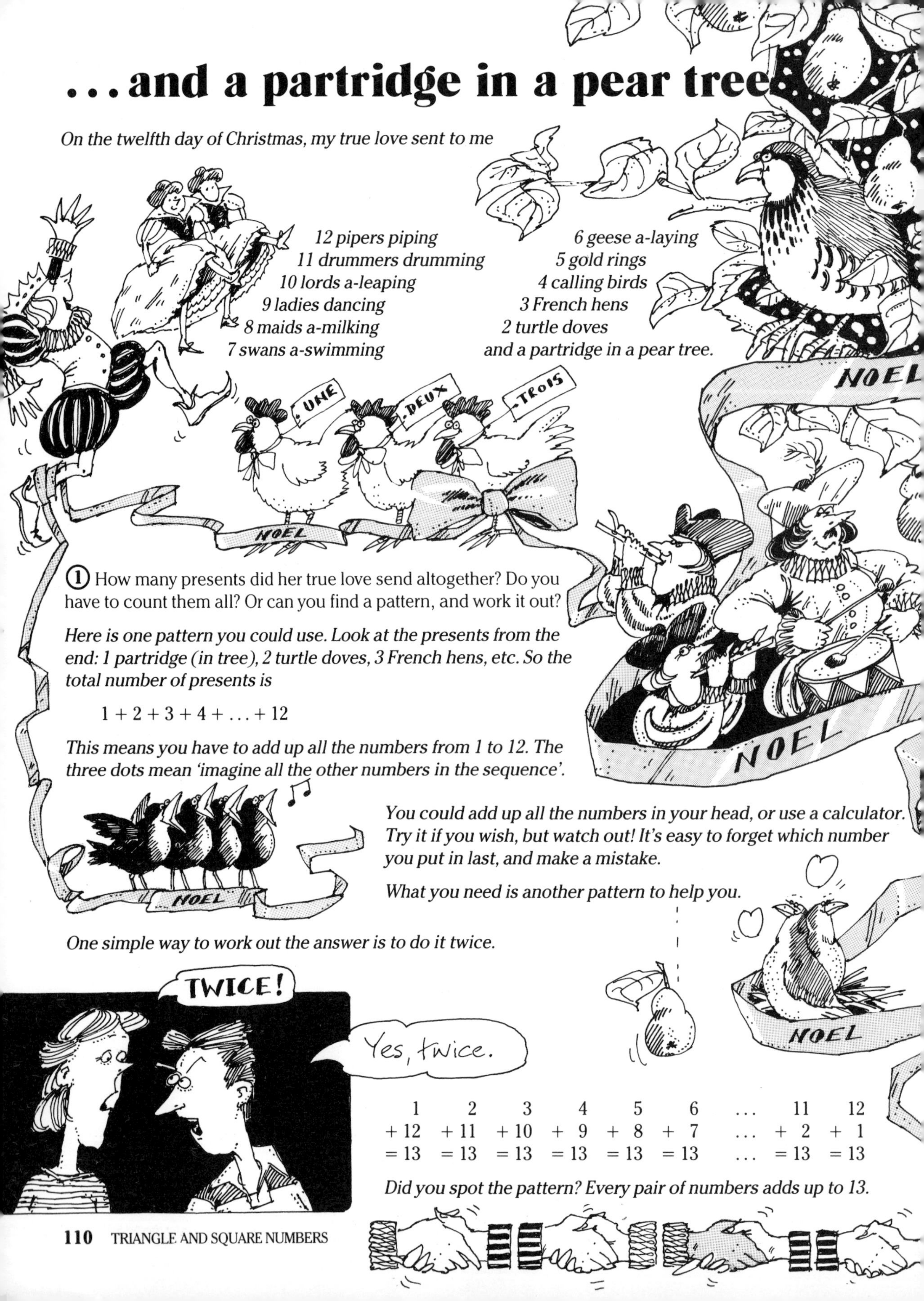

12 pipers piping
11 drummers drumming
10 lords a-leaping
9 ladies dancing
8 maids a-milking
7 swans a-swimming
6 geese a-laying
5 gold rings
4 calling birds
3 French hens
2 turtle doves
and a partridge in a pear tree.

① How many presents did her true love send altogether? Do you have to count them all? Or can you find a pattern, and work it out?

Here is one pattern you could use. Look at the presents from the end: 1 partridge (in tree), 2 turtle doves, 3 French hens, etc. So the total number of presents is

$$1 + 2 + 3 + 4 + \ldots + 12$$

This means you have to add up all the numbers from 1 to 12. The three dots mean 'imagine all the other numbers in the sequence'.

You could add up all the numbers in your head, or use a calculator. Try it if you wish, but watch out! It's easy to forget which number you put in last, and make a mistake.

What you need is another pattern to help you.

One simple way to work out the answer is to do it twice.

1	2	3	4	5	6	. . .	11	12
+ 12	+ 11	+ 10	+ 9	+ 8	+ 7	. . .	+ 2	+ 1
= 13	= 13	= 13	= 13	= 13	= 13	. . .	= 13	= 13

Did you spot the pattern? Every pair of numbers adds up to 13.

Now you can see how adding them up twice will help. There are twelve pairs of numbers, and each pair adds up to 13. So the total of all these pairs is 12 × 13.

But that means you have counted each present twice. So the total number of presents is half of 12 × 13.

Total number of presents from true love is $\frac{1}{2} \times 12 \times 13$
$= 6 \times 13$
$= 78$

A snooker game opens with 15 red balls in a triangle.

② What other numbers of balls could you fit into triangles? Copy these triangle patterns and then fill in the missing ones.

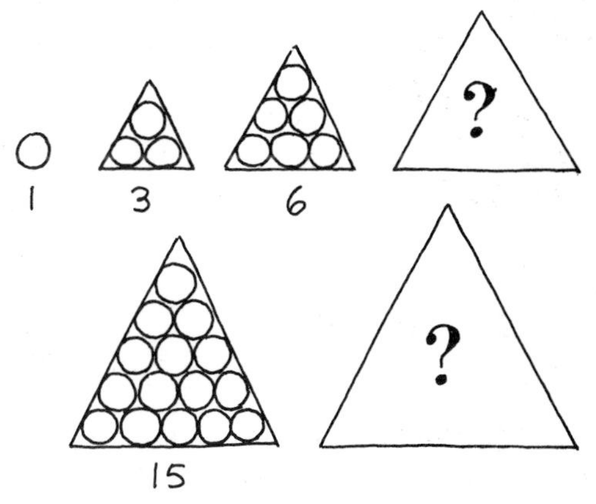

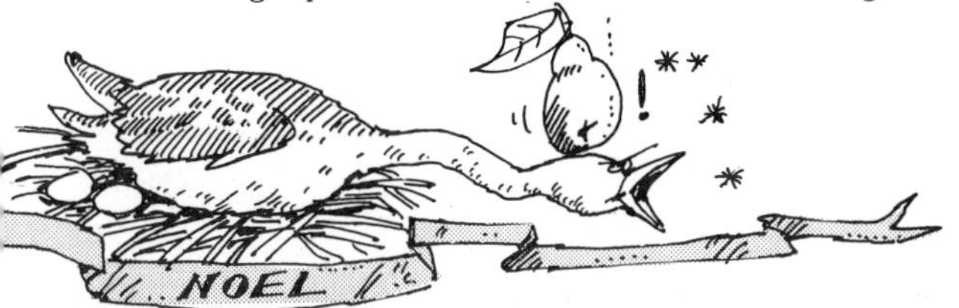

These numbers are called ***triangle numbers****. The first triangle number, △ 1, is 1. △ 2 is 3; △ 3 is 6; and so on.*

③ Write down the difference between the first and the second triangle numbers; then between the second and the third, and so on. Use the pattern you find to work out the seventh, eighth, and ninth triangle numbers.

Triangle number	△ 1	△ 2	△ 3	△ 4	△ 5	△ 6	△ 7	△ 8	△ 9
	1	3	6	?	?	?	?	?	?
Difference		2	3	?	?	?	?	?	?

④ Find out what sort of numbers you get when you add together two next-door triangle numbers; e.g. △ 2 + △ 3, or △ 7 + △ 8.

⑤ A number of people meet at a party. Everyone shakes hands with everyone else. How many handshakes are there altogether when three people meet?

When three more people arrive, are there twice as many shakes? Make a table of the total number of handshakes between various numbers of people, from 2 to 10.

Try out some of them, to make sure.

How many handshakes would there be if the twelve pipers took a break from their piping, and all shook hands with one another, and with the partridge in the pear tree?

Squares

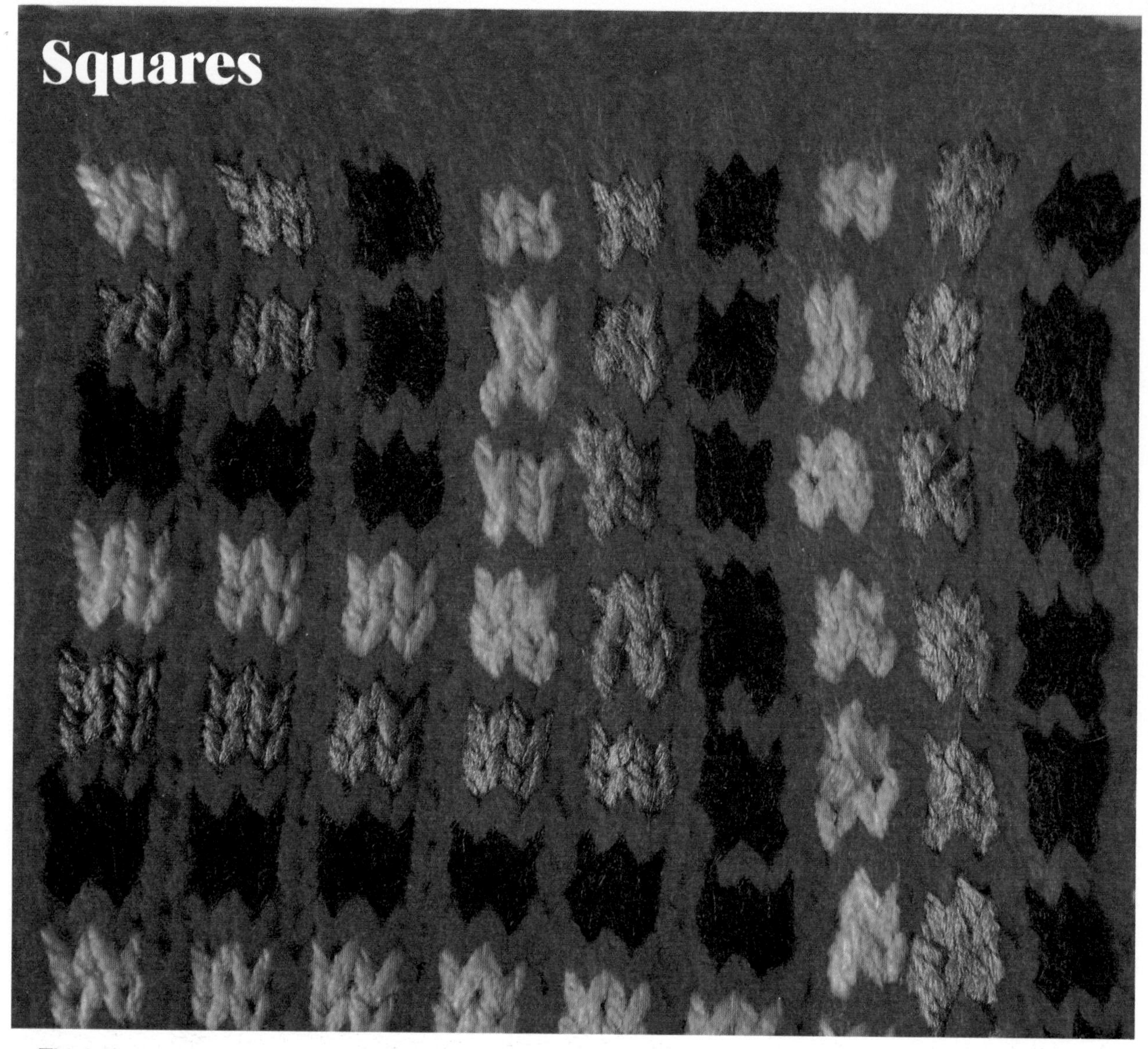

Think like a mathematical moth. You are in the top left corner of this knitting, and you find one yellow square in the red field.

Explore down and to the right. You find three green bits. Together with the first yellow bit, they make a square.

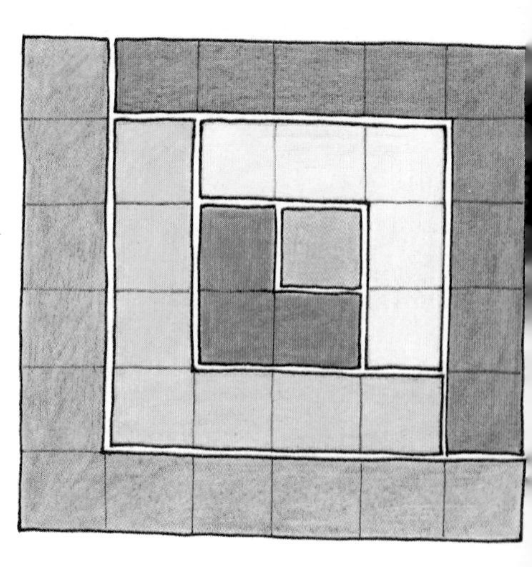

① How many black bits does it take to make the next square?

As you go further down and to the right, you find some more yellow bits; then more green; then more black.

② Write down the numbers of bits in each new band of colour. What sort of numbers are they?

③ In how many different ways could you put L-shapes like these together to make squares?

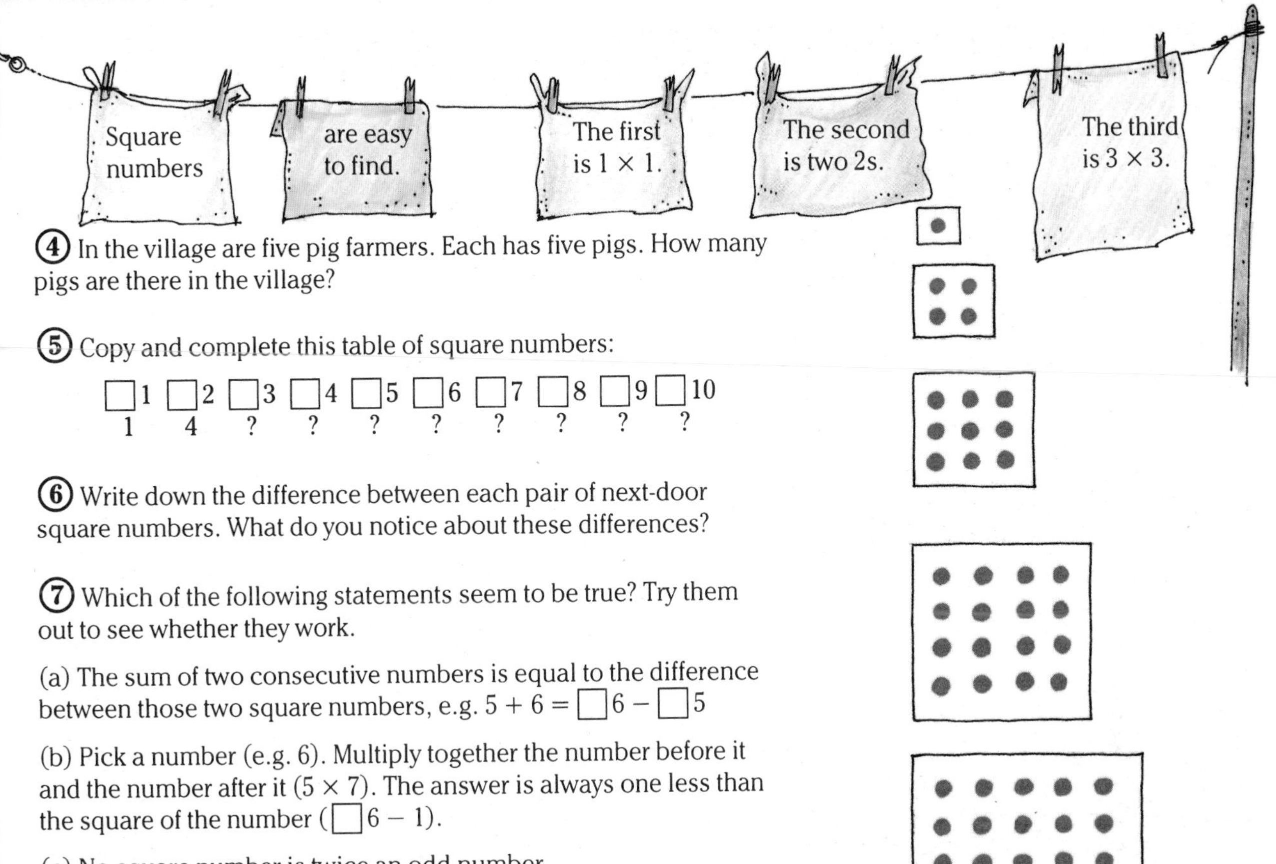

④ In the village are five pig farmers. Each has five pigs. How many pigs are there in the village?

⑤ Copy and complete this table of square numbers:

□1	□2	□3	□4	□5	□6	□7	□8	□9	□10
1	4	?	?	?	?	?	?	?	?

⑥ Write down the difference between each pair of next-door square numbers. What do you notice about these differences?

⑦ Which of the following statements seem to be true? Try them out to see whether they work.

(a) The sum of two consecutive numbers is equal to the difference between those two square numbers, e.g. 5 + 6 = □6 − □5

(b) Pick a number (e.g. 6). Multiply together the number before it and the number after it (5 × 7). The answer is always one less than the square of the number (□6 − 1).

(c) No square number is twice an odd number.

(d) No square number ends with a 2 or a 7.

(e) Subtract 1 from any odd square number and you get eight times a triangle number.

Look at this knitting. Take away the red spot in the centre of the square, and you are left with eight equal triangles, four yellow and four green. Does this help you with (e)?

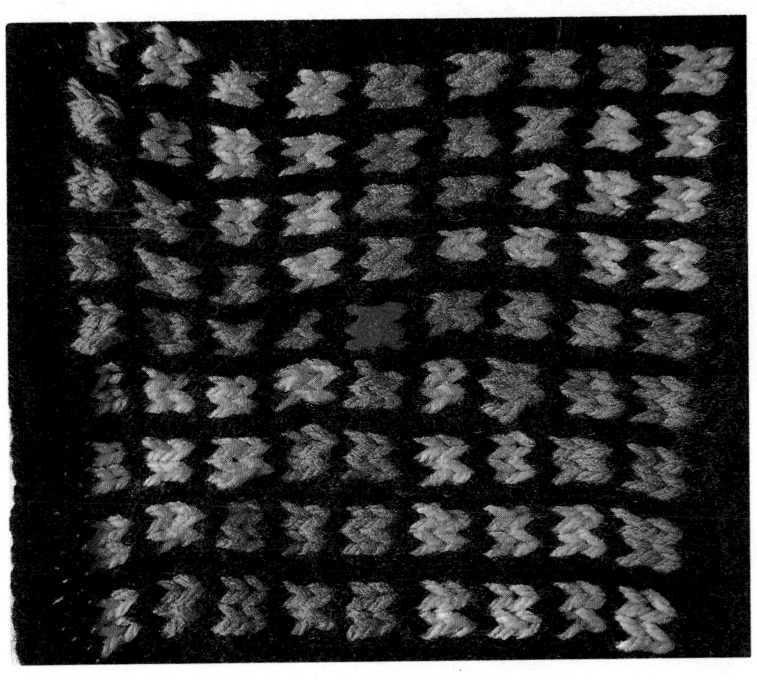

Pascal's triangle

In 1653 a rich nobleman got fed up with losing money in the Paris casino; so he hired a mathematician to tell him the best way to bet. The mathematician's name was Blaise Pascal.

Pascal sorted out the nobleman's problems. He also wrote a book about a triangle of numbers. This triangle was known to the Chinese at least 500 years before Pascal, and it is such a useful pattern that it is still solving problems today.

Each number in the triangle is the sum of the two numbers above it:

$1 + 1 = 2$ $\quad$ $1 + 2 = 3$ $\quad$ $4 + 6 = 10$

① Copy Pascal's triangle into your book. Work out and include the next three rows.

There is a mixture of odd numbers and even numbers in Pascal's triangle. They form lovely patterns, based on triangles.

② Draw triangles round each small group of odd numbers in your Pascal's triangle. Predict with triangles what will happen in the next few rows.

③ What happens if you add up all the numbers across each row? In row 5, for example, $1 + 4 + 6 + 4 + 1 = 16$. What pattern can you find in the sums for the rows?

④ What sort of numbers are these? (Hint: see page 111)

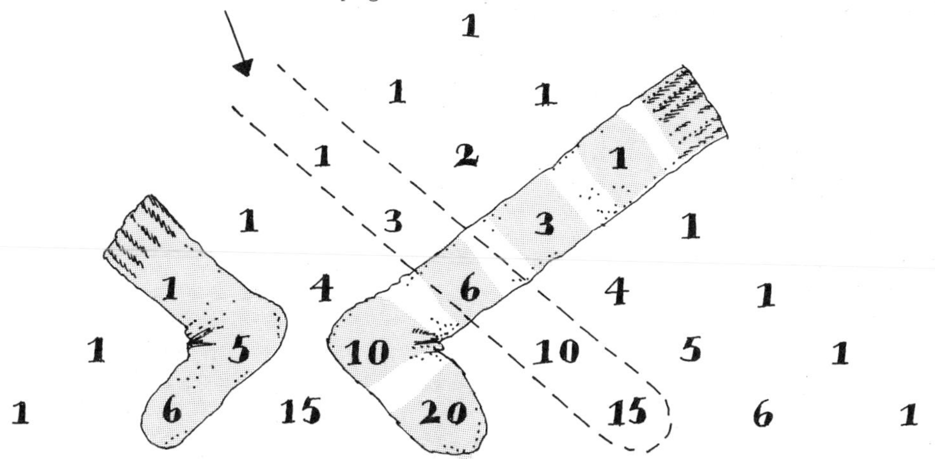

When he went to bed at night Pascal used to hang his socks over his triangle, and one night he noticed a curious thing. The number covered by the toe of a sock always seemed to be equal to the sum of the numbers covered by the leg.

⑤ Look at the left sock above. 6 under the toe, and 1 and 5 under the leg; and $6 = 1 + 5$. Does this work for the sock on the right? Does it always work, wherever you hang a sock?

Imagine that you are the manager of a supermarket. This week you want to sell more tinned strawberries, and you decide to build a big pyramid of cans by the front door, so that the moment customers come in they will think strawberries.

You have to carry all the cans from the store room at the back; so you don't want to take too many. First you decide to try building a triangular pyramid. Each layer will be a triangle. There will be 1 on top, 3 in the next layer, 6 in the next, and so on.

⑥ Write down the number of cans you will need for each of the top six layers. (Hint: see page 111.)

How many cans do you need altogether for the top two layers? Three layers? Four? Find these numbers in a diagonal of Pascal's triangle. Predict how many cans you would need for seven layers.

⑦ Here's another triangle, made like Pascal's. But this one has 2s down one side, instead of 1s.

```
          1
        2   1
      2   3   1
    2   5   4   1
  2   7   9   5   1
2   9  16  14   6   1
```

What sort of numbers are these?

And how would these help you to build a square pyramid?

Investigate some other triangles. Try 2s down both sides. What patterns can you find in the numbers?

The Tower of Hanoi

There is a legend about a monastery in the city of Hanoi. Three diamond needles rise from the floor. The monks have orders from God to move 64 golden discs from one needle to another.

They may not put the discs on the floor; they have to put them on the needles. They may move only one disc at a time, and they may never place a larger disc on a smaller one. When they move the last disc, the world will end with an almighty clap of thunder.

Make yourself a 'tower' of four discs. You can cut the discs from cardboard, making sure that they are all different sizes. Or you can use coins, such as 10p, 2p, 5p, and 1p. Draw three bases on a piece of paper. Label them A, B, and C.

① Try moving the pile from A to B or C. Remember the rules: only one at a time, and never put a big disc on a smaller one.

Here's a useful trick when you are solving problems: ***try working out a simpler example of the same problem first.*** *If you get stuck with four discs, try the puzzle with only three—or only two!*

Another useful trick: ***look for patterns.*** *If you can spot a pattern for a few discs, it may help you work things out for a large number.*

② When you've worked out how to move the discs, write down which way the smallest disc moves (e.g. A–B–C–A . . .). How often does it move? Try to use the movement of the smallest disc to work out a rule for doing the whole puzzle.

You may also be able to find patterns for the movements of the bigger discs.

Over the years, the monks in Hanoi should have learned how to shift the discs in the most efficient way; that is, in the smallest possible number of moves.

③ To shift 1 disc to another pile takes 1 move. Shifting 2 discs to another pile needs 3 moves. Write down the smallest number of moves you need for 3 discs, and for 4 discs. Look for a pattern. Predict how many moves you would need to shift 5 discs to another pile.

Check by making a fith disc for your tower, and counting the moves.

Then predict roughly how many moves you would need to shift six and seven discs.

Sadly, the legend isn't true. There are no monks with golden discs. The Tower of Hanoi was just a tall story made up by M Lucas in 1879 to help sell his puzzle.

But suppose it had been true. Suppose those monks had been moving one disc every second for a million years. Would they have finished yet? Is the world about to end? Should we be quaking in our shoes?

The Tower of Hanoi is a 'recursive' problem. The same thing has to happen again and again. Look at how it works. To move one disc is easy—call the process M1. To move two discs you must carry out M1, shift the second disc, and then do M1 again.

To move three discs, you do M1, move disc 2, do M1, move disc 3, do M1, move disc 2, and do M1 again.

So you have to do M1 once to move 1 disc, twice to move 2 discs, four times to move 3, eight times to move 4, and so on. The number of extra moves doubles every time you add a big disc at the bottom.

④ Copy and complete this table of the number of moves. What is the pattern of the differences?

Number of rings	1		2		3		4		5		6		7	...	10	...	20
Number of moves	1		3		7		15		31		63		127		1023		1 048 575
Difference		2		4		?		??		??		??					

⑤ You can see from this table that a million moves are needed to shift 20 discs. At one second a move, that would take a million seconds. How long is a million seconds?

⑥ Imagine that the poor old monks have moved 63 rings! They're looking forward to a nice cold drink, because they think they have almost finished. Are they right?

No: the monks are wrong! They are less than half way through their task, for they still have to shift the 64th disc, and then move the top 63 all over again. This means doing the 63-puzzle again; and that means doing the 62-puzzle twice, the 61-puzzle 4 times, the 60-puzzle 8 times, and so on, and so on, and so on.

To shift all 64 discs would take 18 446 744 073 709 551 615 moves; nearly 18·5 million million million moves. At one move every second, working day and night, this would take very nearly 600 000 000 000 years; that is six hundred thousand million years.

There's never going to be enough time for anyone actually to do the 64-disc puzzle. The only way that you could possibly complete it is mathematically—in your imagination.

How to solve problems

① The Post Office wants to issue a new £2 book of stamps that will contain both 15p and 20p stamps, but no others. How many of each stamp could there be? Are there other possibilities?

Problems are often difficult to solve. Sometimes you can guess the answer, but more often you have to be systematic. So it helps if you have a general strategy.

THINK	What plan should I try? Is this getting anywhere?
PLAN	Try a simpler example. Look for a pattern.
TRY	Guess an answer. Does it fit the pattern?
CHECK	Does it work? If not, THINK again.

What did you THINK when you read the stamp-book problem above?

You might have made a PLAN to guess how many 20p stamps the book could have.

TRY five, and see whether that works.

CHECK. It doesn't work; so try again.

This sort of approach is called 'trial and error'. Use it sensibly, and trial and error can be an effective way to solve problems of many different kinds.

When Charlemagne became Holy Roman Emperor on Christmas Day AD *800, he hired a Yorkshireman called Alcuin to teach him some maths. Among the problems that Alcuin wrote down, 1200 years ago, was a version of this river-crossing puzzle.*

② You have to get a wolf, a goat, and a cabbage across a river. There is only one canoe, in which you can carry only one thing at a time. You can't leave anything half-way, and the canoe won't go across without you to paddle it.

How can you get them all across, without leaving the goat to eat the cabbage or the wolf to eat the goat?

Don't forget; THINK, PLAN, TRY, CHECK. One plan might be trial and error: what happens if you take the wolf across first? If that doesn't work, try something else.

Another tip: ***be systematic;*** *try to test all the possibilities.*

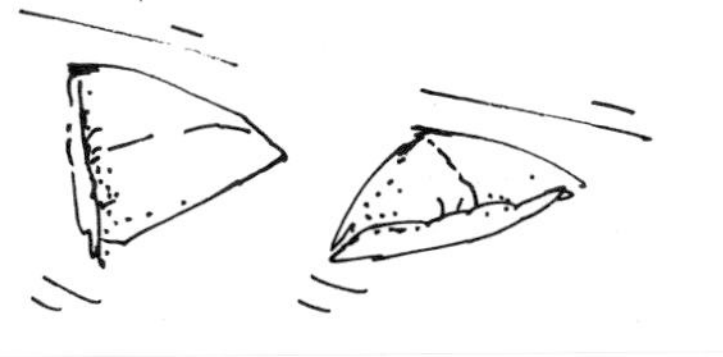

③ Two adults and two children have to cross another river. Their boat will hold one adult or two children, but not one adult and one child. How can they all get across? Remember the boat can't get across on its own; someone has to row it.

④ (a) Before her party, my friend found that if she put four chairs at each table she had one chair left over. If she put five chairs at each table she had one spare table, but no chairs left. How many tables were there, and how many chairs?

(b) She had another problem with the samosas. If she put 7 on each plate, she ran out of plates with 2 samosas left over. But if she put 8 on each plate she had 1 samosa too few to fill all the plates. How many samosas had she made?

⑤ **Loser takes last** is a game you can learn to win. There are two players. They need a pile of counters, or paper-clips, or sweets. Suppose there are 12 in the pile to start with.

When it's your turn, you take either 1 or 2 or 3 from the pile—whichever you like, but you must take at least 1.

The object of the game is **not** to take the last one, but to force the other player to take it.

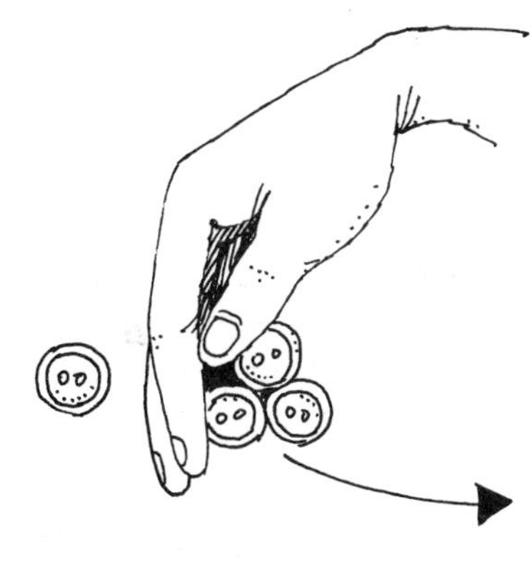

I'm not going to tell you how to win, but you can work out a winning plan by a logical process, starting with A and B.

A. If only 1 is left and it's your turn, you have a losing position. Whoever has to play must take the last one, and lose. So you must make sure you don't get left with 1.

B. If 2 or 3 or 4 are left and it's your turn, you have a winning position. You can take just enough to leave 1.

From here, you can carry on. Work out a table of winning and losing positions throughout the game. When you play, you must try to leave a losing position after each of your turns.

What happens if you change the rules? Instead of taking 1, 2, or 3, try a game in which you can take only 1 or 3 at each turn. You should be able to find a simpler winning plan. But if you can take 1, 2, or 4 at each go, then it's not so easy.

Problems are less of a pain in the neck if you THINK and you PLAN and you TRY and you CHECK.

Teacher's treat

Tick . . . Tick . . .

The hand jerked slowly round the clock above the classroom door.

Dribble . . . Dribble . . .

The rain snaked down the outside of the grimy classroom windows.

Drone . . . Drone . . .

Mr Boring was living up to his name, as usual.

Imagine that your maths teacher was boring. So mind-bendingly dull that just thinking of him puts people to sleep instantly. He could work on hospital radio, as a cure for insomnia!

How would you like to send him away for a trip, all on his own?

Your job is to plan his trip. Work it out step by step, so that you send him as far away as possible. And make sure he doesn't have enough money to hurry back.

① You have £1.25 each from 31 people. Make a rough estimate of how much this is altogether. **Do not use your calculator yet**— remember that it's important to guess first (see page 66).

② Now calculate the total, and check that your estimate was roughly the same.

Mr Boring can travel by taxi, train, or coach. Suppose this table shows how much these normally cost.

	Cost	Distance
Taxi	20p	1 km
Train	£1	10 km
Coach	50p	10 km

③ Work out how far he can go for £1, by each kind of transport. Which will take him furthest, for a given amount of money?

On Wednesdays, train travel is cheaper. The Mid-week Special gives a 20 per cent reduction on the fare for any journey over 100 km.

④ (a) Will Mr Boring be going more than 100 km? If so, he will get a reduction.

(b) How much will the train cost for 10 km on Wednesdays?

(c) How far can he go by rail for £1 on Wednesdays?

(d) Does this make the train cheaper than the coach?

⑤ Now you know how far Mr Boring can go for £1 by taxi, train, or coach, you can work out how far each one will take him if you spend all the money the class collected.

Write down the calculations you have to do.

Estimate the answers first, by doing the sums in your head. Then use your calculator to find the answers.

⑥ Find your own town on a map. To which other towns could you afford to send Mr Boring, by (a) taxi, (b) train, (c) coach? (Hint: A circle might help you here—see page 10.)

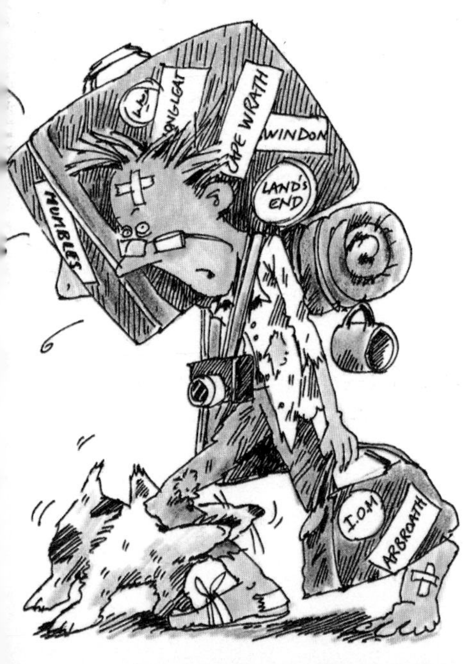

Which is the furthest town you could send him to by coach? When he arrives he will have no money, and nowhere to go. Unfortunately he is going to have to walk home.

⑦ From your map, work out a route for him, and make a note of which places he will have to walk through. If you are feeling kind, you could write a note with instructions, so that he can't get lost.

A good walking speed is about 6 km per hour. But your teacher will also need time to rest, to sleep in haystacks or cowsheds, and to beg for school dinners on the way.

⑧ (a) How many hours of walking will he need to do to get home?

(b) How many days of peace do you think you will get?

Index